ALSO BY JAY INGRAM

The Future of Us

The Science of Why, volumes 1–5

Why Do Onions Make Me Cry?

The End of Memory

Fatal Flaws

Theatre of the Mind

The Science of Pets

JAY INGRAM

Published by Simon & Schuster
NEW YORK   AMSTERDAM/ANTWERP   LONDON
TORONTO   SYDNEY/MELBOURNE   NEW DELHI

SIMON &
SCHUSTER
CANADA

A Division of Simon & Schuster, LLC
166 King Street East, Suite 300
Toronto, Ontario M5A 1J3

For more than 100 years, Simon & Schuster has championed authors and the stories they create. By respecting the copyright of an author's intellectual property, you enable Simon & Schuster and the author to continue publishing exceptional books for years to come. We thank you for supporting the author's copyright by purchasing an authorized edition of this book.

No amount of this book may be reproduced or stored in any format, nor may it be uploaded to any website, database, language-learning model, or other repository, retrieval, or artificial intelligence system without express permission. All rights reserved. Inquiries may be directed to Simon & Schuster, 1230 Avenue of the Americas, New York, NY 10020 or permissions@simonandschuster.com.

Copyright © 2025 by Jay Ingram

All rights reserved, including the right to reproduce this book or portions thereof in any form whatsoever. For information, address Simon & Schuster Canada Subsidiary Rights Department, 166 King Street East, Suite 300, Toronto, Ontario M5A 1J3, Canada.

This Simon & Schuster Canada edition November 2025

SIMON & SCHUSTER CANADA and colophon are trademarks of
Simon & Schuster, LLC

Simon & Schuster strongly believes in freedom of expression and stands against censorship in all its forms. For more information, visit BooksBelong.com.

For information about special discounts for bulk purchases,
please contact Simon & Schuster Special Sales at 1-800-268-3216
or CustomerService@simonandschuster.ca.

Manufactured in the United States of America

1 3 5 7 9 10 8 6 4 2

Online Computer Library Center number: 1492482338

ISBN 978-1-6680-6926-4
ISBN 978-1-6680-6927-1 (ebook)

To Oscar and Crosby, who already live in a world of pets

Contents

Introduction   1

PART I: A BOND LIKE NO OTHER

CHAPTER 1   Biophilia   9

CHAPTER 2   What Is a Pet?   21

CHAPTER 3   When Did It All Start?   27

CHAPTER 4   Why Do We Keep Pets?   31

CHAPTER 5   Does Pet-Keeping Even Make Sense?   38

PART II: THE UNLIKELY VICTORS

CHAPTER 6   Where Did Dogs Come From?   49

CHAPTER 7   Where Did Cats Come From?   59

CHAPTER 8   Costs/Benefits to Humans   69

CHAPTER 9   Costs/Benefits to Pets   78

PART III: SORRY, WE'RE OUT OF DOGS AND CATS, BUT . . .

CHAPTER 10   Horses   87

CHAPTER 11   Parrots   98

CHAPTER 12   Ants   107

CHAPTER 13   Hydras   115

Contents

PART IV: PEOPLE, PETS ... AND MORE PEOPLE

CHAPTER 14	Pet People	125
CHAPTER 15	Pet Names	133
CHAPTER 16	The Whimsy of Dog Breeds	138
CHAPTER 17	Exotic Pets—Exotic People?	147

PART V: DOWNSIDES

CHAPTER 18	The Outdoor Cat	155
CHAPTER 19	Beyond Cats	166
CHAPTER 20	Eat ... or Be Eaten	177

PART VI: ODDITIES

CHAPTER 21	Road Trip	187
CHAPTER 22	A Weird Relationship: The Origin of Dogs, Magnetism, and Poo	192
CHAPTER 23	Do Dogs Know Calculus?	197
CHAPTER 24	Do You Look Like Your Dog?	200
CHAPTER 25	A Psychic Dog	205

PART VII: THINKING LONG-TERM

CHAPTER 26	The Best-Before Date	213
CHAPTER 27	Bring 'Em Back Alive	217
CHAPTER 28	Could There Be Robot Pets?	226

Contents

PART VIII: THE FUTURE

CHAPTER 29	Talk to the Animals	235
CHAPTER 30	The Wild World of Communication	238
CHAPTER 31	You Talkin' to Me?	245
CHAPTER 32	A Cautionary Tail	250
CHAPTER 33	Future Pets	257
	Conclusion	269
	Acknowledgments	273
	Notes	275

Introduction

At the moment, I have one pet, a standard poodle–wheaten terrier mix named Robbie, named after the Canadian-designed screwdriver, the Robertson.* He is something of a mystery to me, but he's not the only pet that's caused me to feel that way. These include a standard poodle (Buster), who defied the odds and lived to seventeen, four cats, two lizards, and two turtles. When I was a kid, we had goldfish—in a bowl—but they had preceded me, so they weren't really *mine*. There was also a dog, Tinker—a spaniel, I believe—who was hit by a car when I was very young. Normally, I never think about this animal. The only shred of a memory I might have is a photo in an album somewhere. It shows the dog together with my older brother, who definitely had more experience with the dog. However, the fact that I don't remember Tinker doesn't mean he didn't influence me: Some research suggests that when a dog licks your face, it might transmit some of the bacteria in its microbiome, which, in turn, might take up residence in your gut.

When I look back at these animals, there's a fuzzy line that divides mammals on one side from reptiles, fish, and amphibians on the other. This will not apply to everyone, but I had two quite different approaches to these pets. I engaged with the cats and dogs,

---

* Long story.

had favorites in both species, and was saddened when they died. The others were different. I liked watching my turtles, red-eared sliders, one of the most common turtle pets in North America. But I didn't bond with them. The lizards were also worth watching, but again, not much serve-and-return. They had occasional bursts of behavior triggered by food, especially active food, like crickets. (Of course, when we get to the point where we understand crickets' mentality much better—and there's definitely research going on in this area—we might come to think this is unacceptable behavior on my part.)

Emotional connections with dogs and cats happen easily. But how do you relate to these other animals? In my case, at least, not emotionally. But they were all fascinating living organisms and well worth having in a home. We just didn't have the same sort of relationship.

There's an implicit note of "foreignness" at work. Not foreign geographically, but in appearance. Yes, an insect has eyes, a mouth, legs, and a heart like we do, but it also has a hard-shelled exoskeleton, *six* legs, compound eyes, and likely can fly. And bite and sting. It definitely has an alien quality compared with, say, a hamster.

But can insects be pets?

There are numerous standard online definitions of a pet along the lines of a domesticated animal kept for companionship and pleasure. Unfortunately, "companionship" and "pleasure" have a lot of flex, and while we all have some mental rules for defining a pet, where did these informal rules come from? Were they shaped by experience or part of some built-in brain software? Regardless, they have become firmly established. When I use the word "rules," it suggests that in most comparisons, it's easy to tell the pet from the non-pet. But I'm not sure about that. Coming up, a list of animals to label "pet" or "non-pet." It's worth trying to keep in mind why you decide one way or the other.

Read this list of animals, and label each "pet" or "non-pet" as you go: a dog, a cat, a hamster, a gerbil, a caged bird, a horse, a pig, a

Introduction

praying mantis. A short list, but already the line separating a pet from a non-pet has been blurred. Yes, some of you will have had a horse that you'd call a pet; some a pig. But horses and pigs are different from dogs and cats. Let's say a piglet becomes very attached to a farmer, and the farmer responds, adopts Piglet. But Piglet's littermates are then routinely "sent to market," so Piglet was the pig who "stayed home." Was Piglet a pet and the others not? What qualified Piglet as a pet?

Some horses and pigs are definitely pets. Many are not. Most horses work for a living, whether at the racetrack, the arena, or the farm. Most cats don't work (yes, they catch mice, but that's freelance activity, not employment). The hired hand/nonworking guest distinction comes into play. Yet there's no doubt that horses and pigs are smart and expressive enough to create an emotional attachment.

In some ways, the definition is revealed in how you describe the interaction you have with a pet. In my case, I talk *to* my dog, but I talked *about* my lizards. I *watched* my lizards—and my turtles—but I *play* with my dog. He's a companion. My lizards and the turtle were not companions.

In the turtle's case, it wasn't his fault. I just didn't see opportunities for any kind of communication at all. The original one, Timothy, might have been aware of me when I stuck my hand over his bowl (the one with the plastic palm tree in the middle), but he paid so little attention it was hard to tell. There were times when he seemed to be eyeing me as I moved around his bowl, but maybe not. But does that make him less of a pet? Even if you don't have a strong emotional link to a pet like that, you could still be fascinated because of, not despite, its foreignness. I'm confident that most would agree that the fact that I fed him and kept him warm in the winter makes him a pet.

One thing is certain: We are learning so much about animal intelligence and animal consciousness that many of our much-loved stereotypical judgments may have to change. What if Timothy was actually smart and just undemonstrative, just kept his thoughts to himself?

Introduction

Maybe if I'd dedicated myself to training him to swim over to greet me, I would have built a stronger relationship.

You really can't be sure, especially when you hear about bumblebees that learn a task by watching more experienced bees, or when you learn how scientists are gaining insights into the grammar of sperm whale conversations, or when you see a video of an orangutan driving a golf cart and you think, "You know, it could be an AI fake, but even if it is, it's totally believable." We have videos of thirsty crows doing exactly what Aesop claimed they could do 2,500 years ago: If the water level in a bottle of drinking water is too low and out of reach, a crow will drop stones into the bottle to raise it.

How broad can the interpretation of "pet" be? Let's extend that list of pets and non-pets a little further. What about invertebrates, like beetles or ants? Stag beetles became a must-have in Japan, triggered by the video game *Mushiking: The King of Beetles*. They don't encourage interaction, but they apparently can be pets. Or an ant colony? An intricate social system that I'm sure you could watch for hours. A pet?

I'm not avoiding the most interesting invertebrate of all, the octopus, but I do *not* support keeping these animals under glass in a home. So I wouldn't advocate for having one as a pet. But what about an alternative? *My Octopus Teacher*, a huge hit on Netflix in 2020, suggests that a relationship can be established if the context is right. Craig Foster was living on the South African coast, taking a break from the rest of his life, and started to document an octopus's life—or, rather, his and the octopus's lives. It provides some fabulous scenes, but to me, the most important thing about a movie like this is that it shows that, with care, it would be possible to have "semi-pets," animals that live in the wild but interact in some way with a person or people. This idea is fraught with negative consequences, especially the well-known ones of habituating wild animals to humans, the losers almost inevitably being the animals.

Take the crow of the fable. You could have a crow as a pet, but what

Introduction

about a somewhat looser arrangement, one in which you attract the crow with a feeder and carry on a relationship—yet the crow stays wild. Of course you can argue that it's bad to make a bird dependent on you for food, but there are something like fifty-seven million homes in the United States that feed the birds one way or the other. It's also worth remembering that Jane Goodall provisioned her chimps when she first arrived at Gombe. If a crow could be a pet with no stay-over privileges, so could the chickadees that will perch on your hand to take a sunflower seed. But if they don't live with you, does that count?

Returning to the original definition, chickadees can't be pets, because they're not domesticated. But then, neither were the lizards nor my turtle. They weren't bred over hundreds of generations to be pets, as dogs and cats have been. Maybe the chickadees don't qualify, because I can't guarantee seeing them when I want to. But I still care about them as if they were pets.

In these cases, there's not much that's inherent in the animal that makes it a pet or not. That label is the human's to adopt or not.

I'm all for the idea of pushing out the boundaries of the typical view of pethood, including bird-watching.

As the world's population continues to move into cities (according to the United Nations, two-thirds of people will live in urban settings by 2050), contact with a wide range of other species will become more and more challenging.

In this book, I will address some fundamental science issues, such as: Why do we keep pets? What's in it for us? For that matter, what's in it for them? How have species that have been domesticated for millennia, such as dogs, cats, and horses, evolved to live with us? How have we literally shaped our companion animals, sometimes for the shallowest reasons? And what will pets of the future look like if we continue to do that?

What kinds of people own pets? Is the owner of a python who walks around with it wrapped around his/her shoulders the same sort

Introduction

of person as a Chihuahua owner? Is it all really the same affection for life simply exhibited in different ways?

The bottom line is always that pet owning is all about the human, not the animal. Whether that will ever change, I doubt, even in the face of the growing understanding that our pets are smarter and more knowledgeable than we ever used to give them credit for. Changing terminology (not "pet" but "animal companion," not "owner" but "guardian") won't do it.[1]

It's important to remember that pet-keeping is a human activity with all the usual contradictions. Yi-Fu Tuan, in his book *Dominance and Affection: The Making of Pets*, put it this way: "Dominance may be cruel and exploitative, with no hint of affection in it. What it produces is the victim. On the other hand, dominance may be combined with affection, and what it produces is the pet."[2]

– PART I –

A Bond Like No Other

− CHAPTER 1 −

Biophilia

In 1984, the renowned biologist Edward O. Wilson, already famed for his studies of ant societies, published a slim book called *Biophilia*.* Subtitled *The Human Bond with Other Species*, the book has had a profound impact on how we think about life on earth. Blending his own world travels with the history of biology and the bond between science and the humanities, Wilson built the case that humans are uniquely attracted to all other species, plant and animal.

He wasn't shy about it either. In the first edition of the book, Wilson made statements that strongly suggest he believed biophilia was innate; that is, genetic: "Biophilia . . . I will be so bold as to define as the innate tendency to focus on life and life processes."[1]

Did he really mean that we have genes that underlie our "focus on life"? If so, they are there because they promoted survival in the past—they've been naturally selected. You don't have to rack your brain to think of what might look like obvious examples: being attentive to the movements of prey like deer, or the calls of the wolf, or knowing through learned experience (tuned by genes) where to find food—plants, insects, other animals—as challenging as that might

---

* The originator of the term was the psychoanalyst Erich Fromm, who defined "biophilia" as "the passionate love of life and of all that is alive" (*The Anatomy of Human Destructiveness* [Henry Holt, 1973]).

have been. Yet the idea of biophilia being a genetic trait seemed an overreach to some, and concrete evidence for that has been hard to come by. Wilson himself changed his stance somewhat nine years later, with this statement: "Biophilia is not a single instinct, but a complex set of learning rules."[2] That either muddied the waters or opened the door to new ideas, depending on your point of view. A surprise twist to this? You could even use the exact opposite—biophobia—to argue that the much better evidence for the genetic basis of fear of spiders and snakes shows that emotional attitudes toward nature, positive *or* negative, can be genetic, as Wilson himself believed: "We need to include biophobia under the broad umbrella of biophilia."[3]

Whether biophilia is indeed genetic or, more likely, some combination of genes, culture, and individual variation, it plays a supremely important role in pet-keeping. "Pet-keeping" is a broad term; you'd expect it to manifest in a myriad of ways, and indeed, there's a huge variety of relationships between people and pets. We all know pet owners who treat their pet as a member of the family, but even E. O. Wilson seems to have treated his pet lizard Methuselah as just another piece of lab equipment on the table: "It would often remain in the same spot for hours or even days without changing its position."[4]

Beyond pets, there's no question that we have a special interest in, and focus on, other forms of life, but sometimes that interest can go a little haywire. Given the interest is expressed by humans, it's not surprising that it can be exclusively focused on humans. Here are two examples, one ancient, one contemporary; one used animals for political purposes, the other for boosting self-image. But first, a brief history of the artistic origins of biophilia.

The inner walls of European caves at Chauvet and Lascaux in France and Altamira in Spain are enlivened by hundreds of vividly colored images of animals and people. The Chauvet cave art is the oldest, estimated to date to about 35,000 years ago. It is Ice Age art. Several of the animals are long gone, like the aurochs, cave lion, cave hyena, and woolly rhino. Altamira, in Spain, is nearly the same age, with its own array of animals, like horses, goats, and bison. Lascaux, an elaborate series of chambers, is more recent, about 20,000 years old, but again devoted to depictions of local wildlife. The discovery of these caves in the nineteenth century led to a radical new appreciation of early modern humans and their culture—these people were clearly obsessed by large animals, some dangerous predators, others potential dinner.

There is actually evidence of a much earlier connection between art and the animals that surrounded these ancient humans. The Neanderthal people were fascinated by birds' feathers and talons. One example is an arrangement of eight white eagle talons that were cut and polished, presumably intended to make a bracelet. That was an incredible 130,000 years ago. That and the much later Ice Age art in France and Spain confirm an all-consuming fascination with wildlife. Exactly what that fascination meant is still uncertain, but the existence of images of what appear to be human-animal hybrids suggests imaginative, visionary thinking about humans, animals, and nature. At the very least, the relationship went beyond hunting or being hunted: ancient biophilia. But change was on its way.

Move forward millennia to the great kingdoms of three, four, or five thousand years ago and our knowledge of what transpired is much more detailed and authenticated, and it doesn't paint a pretty picture. Wildlife continued to play a central role not in the way it did in the lives of ordinary people, which the cave paintings arguably suggest, but for the very rich and very powerful. And the role of the wildlife was integral to that social standing.

There were menageries, similar to zoos in the sense that they were collections of wildlife unfamiliar to the people who would see them, but with a very different purpose from the zoos of today. Hierakonpolis is a good example.

Hierakonpolis was an important city in ancient Egypt about 5,500 years ago, long before the reigns of the pharaohs. A series of archaeological excavations have revealed what has been called the first zoo, including the bones of baboons, elephants, hippos, wildcats, crocodiles, and a hartebeest—even some dogs. The animals were buried beside some of the most celebrated humans of the time, and there might have been a connection: that is, that the death of a ruler prompted the sacrifice of exotic animals as tribute. If so, this wouldn't be the last time the demonstration of power and influence would be realized by the death of exotic animals. At the same time, though, there was respect; one elephant was buried on a reed mat and covered with linen.

The Egyptian fascination with wild animals continued through the following millennia. Most of the later pharaohs maintained gardens and zoos with animals mostly imported from far up the Nile River. (In fact, recent DNA analysis of the remains of baboons buried in ancient Egypt has located the formerly mysterious Land of Punt as being in or near modern-day Eritrea.) It might not be a step too far to make a connection between the Ice Age paintings of large animals on cave walls to their burial with powerful humans thousands of years later; they might be comparable gestures of respect, awe, and control. We don't really know, but of those three, control soon took over.

In ancient Assyria and Babylon, a mix of conquest, wealth, the need to promote and celebrate a leader's power, and the ability to do that by seizing foreign territory, animals, and human labor conspired to make possible jaw-dropping exhibitions of wild animals.

It wasn't enough just to display the animals. Yes, bringing back a variety of species of deer implied the king's dominance over foreign lands, but it would be even better if he dominated the animals themselves. The Assyrian king Ashurnasirpal II once bragged that, at the behest of the gods, he killed 30 elephants, 257 wild oxen, and 370 lions. These likely weren't what you'd call free-range lions, though.

A series of carved reliefs found in the excavated ruins of the ancient city of Nineveh detailed the hunting prowess of the last king of Assyria, Ashurbanipal. Lions were released from cages in the game park, where he was positioned nearby to thrust a sword into them, or fire arrows from close range. If an arrow missed, he was immediately protected by guardsmen who were standing by. The killing was the point, and the drama associated with it, not necessarily the skill deployed in doing so.

However—and this hints at a deeper relationship between humans and animals—bloodshed wasn't always wrapped into the spectacular display of foreign beasts. A perfect example was the extravagant all-day, mile-long (1.6 kilometer) procession in Alexandria, Egypt, in the middle of February, 278 BCE, a few hundred years after the fall of Assyria. It was a tribute to the god Dionysus, but equally a celebration of the reign of King Ptolemy II. There was a seemingly endless series of chariots pulled first by teams of four elephants apiece, then by antelopes and eight pairs of ostriches (!); buffalo followed, then zebras, spice-bearing camels, hundreds of sheep, cages of parrots, pheasants, peacocks and guinea fowl, Ethiopian oxen, a gigantic white bear, leopards, panthers, lynx, innumerable horses, twenty-four large lions, and a rhinoceros. No word on their final fate.[5]

Somehow, as time passed and Rome became ascendant, violence took center stage again as animals became objects of slaughter in front of huge, approving crowds. While there had been mass killings of animals for joyful Roman crowds before the reign of Emperor Augustus (63 BCE–14 CE), he really kick-started the idea, claiming that he had been responsible, over a total of twenty-six public shows, for the killing of 3,500 animals, including such exotics spectacles as the execution of thirty-six Egyptian crocodiles.

Augustus's alleged kill of 3,500 might have been a solid start, but it was quickly bested by a well-documented 11,000 in a massive 123-day death carnival sponsored by Emperor Trajan in 108–109 CE. Eventually, though, this lust for blood, both human and animal, waned in Rome, ending this bizarre, millennia-long extravagance of animal life and death.

But while the slaughter abated, the attraction of exotic animal collections didn't die out, suggesting that somewhere in there a thread of biophilia remained, manifesting in something closer to the modern zoo, a collection of animals from far-off places to entertain and educate the local crowds. The first zoos of this kind appeared in Europe in the mid-1700s, first in Vienna, then Madrid, then Paris. They were designed pretty much exclusively for the audience, not the animals themselves, and reversing that has been a slow process. Today, zoos make efforts to accommodate their animals in more naturalistic circumstances and restrict themselves to animals born in captivity. Forward-looking zoos have captive breeding programs to reverse, even in some small way, the decline of wild populations. Zoo enclosures are much more natural, and behavior can be studied scientifically.

While these early zoos were the mainstream expression of the human desire for animal menageries, the void left by past Assyrian kings and Roman emperors was occupied by a few who still wished to use collections of exotic animals to project wealth and power.

William Randolph Hearst was a good example. He gradually assembled a one-thousand-square-kilometer (250,000 acre) ranch near San Simeon, California. He built Hearst Castle on the property but also, in the 1920s, began to establish a giant zoo, called the Hearst Garden of Comparative Zoology. Guests drove through it on their way to the castle. The biggest private zoo in the world, it had an open area for herbivores to graze, and enclosures for others. There were about three hundred herbivores, including African and Asian antelope, sambar deer, giraffes, zebras, camels, red deer, axis deer, llamas, kangaroos, ostriches, emus, Barbary sheep, bighorn sheep, musk oxen, and yaks. Carnivores like grizzlies, black bears, tigers, leopards, jaguars, and cougars were kept in the enclosures. When Hearst ran into financial trouble in the late 1930s, the zoo began to sell off the animals, but that process was never completed; even today, you can sometimes see zebras descended from the original ones running wild on the hills near San Simeon.

Biophilia

The Hearst Garden of Comparative Zoology, while obviously a symbol of power, was at least there to entertain the public, if only those on Hearst's invitation list. Cocaine magnate Pablo Escobar's zoo made it halfway there: It was definitely an expression of power, but visiting it was another matter, considering it was at his headquarters, Hacienda Nápoles, east of Medellín, Colombia. The entire estate was not nearly as large as Hearst's, covering about twenty square kilometers (nearly five thousand acres), and the array of animals was smaller, although not too bad for a private zoo: giraffes, zebras, camels, antelopes, elephants, flamingos, rhinoceroses, and, crucially, four hippopotamuses.

As is well known now, after Escobar's death in 1993, the Colombian government dismantled the zoo, transferring most of the animals to other parks or zoos, but the four hippos, three females and a male, were too difficult to handle, so they were released into the jungle. They managed incredibly well, so much so that there is now an estimated population of about two hundred. This is an invasive species, the largest in the world, one that is dangerous (hippos kill five hundred people a year in Africa) and threatens to upset the ecology in and around the Magdalena River. The difficult choice facing the Colombian government is what to do with Escobar's cocaine hippos. Culling is distasteful; sterilizing awkward, dangerous, and expensive; transferring them to zoos complicated and also expensive. But probably the worst option is waiting.

Hearst and Escobar are dead. So is the tradition of the rich and powerful creating a zoo as a monument to self-glorification over? Perhaps not.

Anant Ambani, son of the richest person in India, Mukesh Ambani, the twelfth richest man in the world, has created a giant combination wildlife rescue rehabilitation center and zoo called Greens Zoological Rescue and Rehabilitation Kingdom. Anant Ambani refers to it as an "animal shelter." It has a pretty cool Instagram account (@vantara), and depending on what you read, you find variable accounts of the number of animals either already in residence or pending arrival. It could be nearly five thousand; it could include three thousand herbivores and thirty-three big cats. A spokesperson claimed that no wildlife would be "exposed for entertainment" for his guests, and that safaris would be "solely for educational purposes."[6] Just to tie everything together, the Greens Zoological Rescue and Rehabilitation Kingdom has apparently applied to import cocaine hippos from Colombia.

Even if this becomes a giant personal zoo, albeit with some rescue features, the ancient habit of procuring animals to show them on a grand scale has virtually disappeared. On a grand scale, yes, but that doesn't prevent the existence of Tiger King or people like him who love to have a few exotic animals in their possession. Mike Tyson's Bengal tigers, Paris Hilton's kinkajou, all pretty small stuff really.

Is all this a shade of biophilia? Not really for the human owners—they're in it for their image. But they realized the human-animal link is the most efficient way of impressing the crowds via *their* biophilia.

From the ancient Assyrians and Egyptians on, animals were the tools of the expression of power, control, and wealth. While the modern private zookeepers have eliminated ritual slaughter, it still carries on in another realm.

Here's one detailed account of the end of a trophy hunt:

> My bullet has smashed its shoulder, blowing away the top part of its heart and destroying a lung. Van Aswegen's bullet has broken the spine. He and I circle the lion, then I put two more bullets through its backbone and into the lungs from behind and above. . . . Its boiled-out skull should rank very high in the SCI [Safari Club International, a trophy-hunting advocacy organization] Record Book of Trophy Animals.[7]

These accounts are hard to read but not hard to find. If you're unfamiliar with trophy hunting, it is hunting and killing wild animals to display the slain animal, or its head, or its antlers. Typical targets might be leopards, lions, bighorn sheep, or elephants. It is an expensive activity.

Trophy hunting causes pain and suffering. But it generates a lot of money. Some of that money is said to enhance conservation, although that position is highly controversial.*

And, before going further, I am not talking here about hunting for food, hunting to experience nature, hunting for spiritual or ceremonial reasons, or government-approved hunting of animals deemed as pests. Trophy hunting overlaps with some of those, but its themes are adventure, the chase, reputation, and ultimately personal experience.

I have been in the home of a trophy hunter, an overwhelming display of mounted animals, birds, and fish, including some more elaborate stagings, like elephant-foot umbrella holders and deer-foot lamp stands. There were hundreds. Trophies replaced art and furniture. Surrounding yourself with animal trophies is extreme, though perverse, biophilia.

Of course trophy hunting has nothing to do with pet owning—or does it? I think it does—trophy hunters aren't bragging about their skill on the firing range. An animal, preferably one that's large and fierce, has to be a part of it. Accounts like the one above can be found where the gory details are omitted and the hunter describes the prolonged search, the affinity with the animal, and his/her intense emotional response both to the hunt and the kill. Love of nature bridges the gap between trophy hunters and pet owners, however faintly. We all share this positive feeling and express it in different ways.

---

* Trophy-hunting supporters (which surprisingly include organizations like the International Union for Conservation of Nature) argue that the money hunters spend to bag a giraffe or elephant supports local people and communities, preserves habitat, and ensures that conservation efforts will be sustained. Opponents contend that there's no solid evidence for such benefits and suspect most of the money gets into the hands of less-deserving recipients.

E. O. Wilson, a dedicated conservationist, would never approve of trophy hunting were he alive. Such experiences were remote from Wilson's life as a biologist, walking through forests, ever alert for familiar—or even new—species of ants—or anything! Yet a close reading of Wilson suggests he glimpsed common ground between the naturalist and the hunter when it came to finding the desired species: "The naturalist is a civilized hunter. He goes alone into a field or woodland and closes his mind to everything but the time and place. . . . The hunter-in-naturalist knows that he does not know what is going to happen."[8]

Here is hunter Jadine Jedresko with her attitude toward the hunt: "I have a huge respect for each species that I hunt. . . . You take a picture because it's a respect to that animal. Hunting is not about killing for me."[9]

Taking pictures of an animal you've just killed as a sign of respect is an odd pairing, yet naturalists and trophy hunters do share some similar thoughts, especially the sense of achievement. If you think I'm stretching this, don't forget that old-time naturalists, even Wilson, killed animals or birds if they suspected an undiscovered new species; a museum collection would be their destination. Wilson again: "My Papuan guides stopped hunting alpine wallabies with dogs and arrows, I stopped putting beetles and frogs in bottles of alcohol, and together we scanned the panoramic view."[10]

While it's beetles, frogs, and salamanders in this case, a century before it could have been something on the trophy-hunting scale (where do the animals in old-style museum dioramas come from?). But while some of that experience might be shared with hunters, the purpose is definitely not. Naturalists have their eye on both present and future, accumulating knowledge of what lives where, and preserving that record for the naturalists to come. Trophy hunters want to achieve bigger and more impressive kills, be photographed with them (posed, really), and the bigger the kill, the better. The size is both advertised by hunters' photos on social media and, if it's big enough, acknowledged by one of the hunting organizations that keep such metrics (see the quote from the trophy hunter earlier in this chapter).

So yes, biophilia runs through both the naturalists' and trophy hunters' experiences, but the goals couldn't be more different. Both in a sense are extreme: Few people go to the lengths naturalists like E. O. Wilson and trophy hunters do to achieve their ends, but even as extremes they have something in common.

Pet owning is a strong expression of biophilia. It is probably the easiest too (other than couch-bound adventuring with David Attenborough), but bird-watching, whale-watching, gardening, visiting zoos, and collecting wildlife art would all qualify. But how far does biophilia stretch? Greyhound racing? Rodeo? Other sorts of hunting? People who participate in these activities argue that they respect, honor, and even love the animals involved, despite the obvious downsides for those animals. Can that really be called an outlying expression of biophilia? Does this really connect at all to pets?

The funny thing is that pets have been there all along; while Assyrian kings were killing lions at close quarters, dogs and cats were living quieter lives nearby. When President Teddy Roosevelt was gunning down big game, caged birds and goldfish were kept around the world. When the Hearst Garden of Comparative Zoology was thriving in the 1930s, dogs and cats were cementing their status as the most common animals living with people. Now, with pet numbers exploding, and increasing pressure on the illegal animal trade, with zoos committing to taking no more animals from the wild, pets are ascendant.

There are now several studies showing that pet ownership correlates with positive feeling about animals other than pets, such as farm and wild animals, as well as with pro-environment attitudes and even reduced eating of meat. Owning a pet encourages a closer and healthier relationship between humans and animals.[11]

A Bond Like No Other

Whatever value you would put on interactions with animals—of any kind—as we look forward to 2050, when two-thirds of all humans will be living in cities, you can envision how important pets will be to maintaining the human-animal bond. Animals as ambassadors, not targets.

– CHAPTER 2 –

What Is a Pet?

> A considerable body of evidence indicates that affection for animals is an intrinsic part of human nature that has shaped both who we are today and how we got here.
>
> —John Bradshaw

In the introduction, I listed some animals and challenged you to decide on the spur of the moment which were pets and which were not. Let me now expand that list and try it again: an ostrich, an ocelot, a cocker spaniel, a white rat, a badger, a Bengal cat, a wolverine, a grackle, a lamprey eel, a ball python, a dung beetle, and a white-tailed deer.

A cocker spaniel, a white rat, and a Bengal cat are certainly pets. I once met a guy who had a wolverine living under the front steps that would come into his house and eat the dog's food, but I've heard of no others. (I was standing inside the house while the wolverine was leaving, and I reached down and dragged my hand along its back as it passed by, allowing me to claim forever that I've petted a wolverine.) Ostriches have the disadvantage of being aggressive—especially during mating season—and requiring two-meter (six foot) fencing. But a ball python? There are lots of people who own one of those. You see them on the subway all the time. Dung beetles not so much (how do you feed them?), but stag beetles are a big hit in Japan.

It's not that easy to come up with a strict definition of a pet.

A Bond Like No Other

Online dictionaries coalesce around a small number of criteria: a domesticated animal, kept for pleasure or company rather than utility, and cared for. That covers most definitions, although affection *for* the pet is sometimes added, and Wikipedia clarifies the "rather than utility" argument by defining "a working animal, livestock or a laboratory animal" as a non-pet. However, this is not to say that a herding dog, a dairy cow, or a lab rat *couldn't* become someone's pet—it's just that they usually don't.

Animals that are wild, even if adopted and cared for, seem not to qualify because they're not domesticated; that is, their offspring would not be tame but wild. As suggested above, animals that are provided shelter and fed, but that work for their humans, also fail to qualify—at least by these definitions. It's also obvious that many animals would cross the lines between these definitions.

Katherine "Kasey" Grier, author of *Pets in America: A History*, expands on these definitions in a set of nine features of pets on her website.[1] They're worth reading, but I've briefly summarized them here.

The pet animal is chosen by one or more people who take responsibility for its well-being. It is kept in close proximity to those people, to whom it is a companion. It may be a source of pleasure, leisure activity, or social status. Grier goes on to argue that being a pet isn't a permanent designation. In what might be a controversial point of view, she argues that "the status of the pet is contingent. An animal can be raised into the status of the 'pet' or lose that status through abandonment or other failures of care."[2]

So what do you think? Is a puppy born to a feral mother a pet? Potentially yes, but it would have to be adopted to become one. Living feral from birth to death wouldn't qualify by Grier's criteria, and I agree with her.

Is a dog who belonged to a family but was abandoned and is now living unclaimed in a rescue facility still a pet? It is still being cared for, although perhaps not in an intimate, pet-like fashion, but I feel

What Is a Pet?

it's still a pet, perhaps in the same way a retired athlete still carries the reputation he/she earned when active. Nonetheless, if that same dog ran away and lived the rest of its life in the wild, it would be hard to call it a pet. A wild dog maybe, but not a pet. A *former* pet.

Even this brief foray into what defines a pet makes one thing very clear: An animal on its own is not a pet. All definitions require the permanent presence of a human. An animal not cared for by a human is not a pet. Wild animals, even ones whose lives are intertwined with humans, like the chimpanzees at Jane Goodall's Gombe, are not pets. The raccoon with a broken leg that is rehabilitated by animal rescue is not a pet.

It hardly seems necessary to have these definitions, because most of us know intuitively what a pet is, but it is important to underline the central role of humans because the science of pets is actually the science of humans' relations with those animals we call pets. A pet needs those connections with a human mentioned above to *be* a pet. How did wolves become dogs? Their interactions with humans. What about the chickadees that come to my birdfeeder and take a sunflower seed from my grandson's outstretched hand? They are habituated, not domesticated, so they're not pets, but I feel affection for them. I doubt that affection is reciprocated, though, because if I stop putting sunflower seeds in the feeder, they will go elsewhere, and aside from checking out the feeder occasionally, I'm sure they won't give me another thought. There's also identity—I like *all* the chickadees equally, although I'm confident that every one of them that I was fond of three years ago is likely dead.

Katherine Grier's list of pet criteria includes this: "The pet animal is kept as a companion or the object of emotional ties." My chickadees fail the "kept as a companion" part of that, though they satisfy the "object of emotional ties."

John Bradshaw, in his book *The Animals Among Us: The New Science of Anthrozoology*, agrees: "The animals we loosely refer to as 'pets' may still be many things beside companions. They may be equally valued for their physical beauty (show dogs), for their usefulness (gun dogs), or for their ability in sport (agility champions). . . . But all true pets have one thing in common: the affection of at least one person, who recognizes and values them for their qualities as individuals."[3]

There's that requirement again for both affection and companionship. How far would companionship extend?

In an article in the journal *Anthrozoös* in 2003, Timothy Eddy pointed out that "pet" can even describe certain humans, as in "a human who is treated with unusual kindness or consideration," like a teacher's pet. He then argued that the same definition could perhaps be applied to animals too: that is, a pet as "an animal who is treated with unusual kindness or consideration." No other constraints. As Eddy then pointed out, many animals would then legitimately qualify to be called pets. The unclaimed dog in the rescue facility would still be a pet; my chickadees would be pets. Even animals protected by conservation edicts ("Please Slow Down—Wildlife Crossing") could be considered pets.[4]

It's an approach I like because it allows animals of which we are fond and protective to remain living their independent lives while still benefiting from our concern and generosity.

However, I'm afraid this might be a step too far for most. After all, in his book, Bradshaw makes a telling point: "Nowadays, eight out of ten pet owners consider their pets not just part of the family but as having equal status with the human members." So much for my turtle, the two lizards, and definitely the ant colony. Bradshaw's book was published in 2017; a Pew Research Center report on this subject in 2023 established that 97 percent of American pet owners considered their pets "to be a part of the family."[5] This broke down to (roughly) 51 percent declaring "as much as a human member," but (roughly) 47 percent backing away slightly by saying "yes, but *not* as much as a human member." Even with the equivocation, it's a powerful comment

What Is a Pet?

on the status of both the modern pet and human members of pet-owning families.*

As far as I'm concerned, the emotional attachment is the strongest bond between human and pet, followed closely by—and connected to—a mental bond, a feeling that you have an idea of what your pet is thinking or why it's doing something. I separate those two because your pet might well be doing something *without* thinking. And really, you likely have no idea what your pet is thinking. But as humans, the more we persuade ourselves that we can follow the brain workings of a pet, the more engaged we are.

Pet ownership of course isn't limited to these two factors. Some pet owners sport their pets as fashion accoutrements, others as not-so-subtle statements of aggression. As interesting as these relationships might be, they're outliers when it comes to the central reasons for pet ownership.

Are Humans the Only Species That Keeps Pets?

If pets are tied to humans, if you can't have one without the other, are we the only species that has pets? You can find claims online that there are other "associations" between two different species that might remind you of a pet relationship, but they're not. Koko, the famous captive gorilla in California, bonded with a kitten she called All Ball and was apparently delighted with the kitten's company, but it's not legitimate to characterize a relationship between an actual pet and a captive animal as something that might occur in the wild.

---

* In an earlier Pew Research Center survey, they went so far as to compare how positively American adults felt about not just their dogs and cats, but also their mothers and fathers. Here are the rankings: dog, 94 percent; Mom, 87 percent; cat, 84 percent; Dad, 74 percent.

A Bond Like No Other

A house cat and a wild gorilla would never encounter each other. Most of the other examples that have become popular have the same shortcoming: They are not wild animals, but have come together in some sort of refuge, an entirely unnatural situation. Tarra the elephant and her dog Bella, Amy the deer and her dog Ransom, Tonda the orangutan and her cat TK (short for "Tonda's Kitty")—all were living in artificial environments. And if you see a chimpanzee playing with a monkey in the wild and think "Look, it has a pet," you are obliged to stay to the end of the scene when the chimp bashes the monkey's head against a branch and eats it.

My favorite anthrozoologist (a researcher who studies the relationships between humans and animals), Harold "Hal" Herzog, had stated publicly that he believed there had never been a case of a wild animal taking a different species as a pet, but then, in 2004, he was contacted about an apparent exception to his rule. In Brazil, a band of capuchin monkeys was seen taking care of a marmoset, a different species. This was not a brief interlude either: The capuchins, primarily two different females, took the young marmoset (the biologists tracking the situation called her Fortunata) with them everywhere they went, fed her, made sure that she was safe as they moved quickly through the trees, and generally treated her as if she were a capuchin herself.

Herzog did mention that this too was not a 100 percent "wild" situation, given that these monkeys were living in a biological reserve and were provisioned with food, but you could tell his objections were tempered by the realization that this was as close as you could get to a natural setting. He pretty much had to rely on the admission that "Fortunata may be the exception that proves the more general rule that non-human animals don't keep pets."

This discussion about why so few—if any—other animals keep pets raises the question: Why do we? What's in it for us? It turns out to be a complicated question with roots in evolution, extending far out to human health, both physical and mental.

– CHAPTER 3 –

When Did It All Start?

We have no real idea when our ancestors started to keep pets. As I've suggested, if current theories of how wolves eventually became dogs are true, pet-keeping might have begun thirty thousand years ago, but that is the foggiest image of an "origin" you could imagine. The uncertainty here is the direct result of lack of tangible evidence. And even when there is evidence, placing it in the historical context is a challenge. For instance, a grave in northern Israel that is at least 12,000 years old contains the remains of two individuals: One was an adult human whose sex is indeterminate because the pelvis was crushed; the other was a puppy that died at probably a few months old. It's their postures that are intriguing. The human lies on his/her right side, head resting on the left wrist, while the left hand is draped over the puppy's bones. It would be difficult to depict a close—probably emotional—relationship more vividly.[1]

A human, a dog—early traces of pet-keeping? Maybe, but in the same area, there is a mass grave from a few thousand years earlier containing the remains of humans and a variety of animals, including deer, gazelles, and tortoises; and one grave with a fox and a human decorated with red ocher dye. A human buried with a fox complicates the picture because there is no evidence that foxes were ever domesticated (at least until the twentieth century; see chapter 6). There is also an abundance of the red ocher in this particular burial, usually a sign of some ceremonial interest. Does this site suggest some tentative moves toward domestication, or is the mini-zoo of animals buried with humans indicative of some other, as yet unidentified, trend?

The archaeologists who unearthed these graves argue that the fact the fox skull had been previously interred in a different grave and was subsequently moved in with the human was significant; also, while fox bones appear in many burials of the time, they usually bear signs of having been butchered or consumed. These fox remains do not.

For context, the people of the time of the mystery fox are usually said to be in transition, not yet growing crops, still hunting and gathering, but gradually settling into villages. It's only when village life took over and hunting and gathering were abandoned that domestication of both animals and plants began to happen. But that wasn't until about ten thousand years ago—these examples of animal-human burials predate that. It might be that these examples were part of a run-up to domestication, but evidence for what societies were doing pet-wise in the thousands of years prior to this is extremely rare.

However, it is possible to make some guesses informed by observing modern people who do not live in Western urban societies. Cultural anthropologist Loretta Cormier at the University of Alabama has studied a variety of groups living along the Amazon River and points out that pet-keeping is common. In fifteen months with the Guajá people, she observed ninety monkeys being kept as pets in a village of about one hundred people. But there is a crucial difference between the Guajá and most modern pet-keeping: Cormier points out that the same animals that are kept as pets are also hunted for food. When an adult with young is killed, some of the young are adopted as pets and escape becoming lunch. The rest do not. Most of the time, once an animal is designated as "pet," it is not eaten, even when an adult.

The Guajá told Cormier the only animals they wouldn't consider adopting as pets were snakes.[2] In the Guajá, men and women play different roles in pet-keeping. When a family group of animals is

When Did It All Start?

captured, and adults and young are separated, the young aren't automatically designated as pets. The wives of the successful hunters make the choice. The lucky ones that do become pets are the responsibility of the women, and each pet's caretaker is considered its "mother." They believe women have a way of understanding pets that men lack, and that animals share that belief. (Some studies of men and women in Western pet-keeping practices are similar, as I discuss later.)

One of the most intriguing instances of pet-keeping was in Tahiti. In a book published in 1841 called *Polynesian Researches*, the author, William Ellis, describes pet eels:

> Eels being great favourites, are sometimes tamed, and fed till they attain an enormous size. Taaroarii had several in different parts of the island. These pets were kept in large holes, two or three feet deep, partially filled with water. On the sides of these pits, the eels formed or found an aperture in a horizontal direction, in which they generally remained, excepting when called by the person who fed them. I have been several times with the young chief, when he has sat down by the side of the hole, and, by giving a shrill sort of whistle, has brought out an enormous eel, which has moved about the surface of the water, and eaten with confidence out of his master's hand.[3]

This is not a practice limited to monkeys, eels, or indeed the tropics. The Selkup people of Siberia are said to have captured the puppies of Arctic foxes. They would take them shortly after birth in the springtime, feed them, keep them in cages, but then, unlike the Guajá, rather than keep them as companion animals, would slaughter them for food in the fall.[4]

My favorite Siberian example concerns the ermine, the variety of weasel that turns white in the winter. They too have been commonly kept as pets, although that might not be quite the right word, as they retain a certain wildness. Veronika Simonova, a social anthropologist, quoted a man in Siberia as saying that "ermines visit people quite often in log cabins or tents in the forest."[5]

That *exact same thing* happened to me! I was alone in my place in the woods north of Toronto when a white-coated ermine suddenly appeared in the living room, scampering around, checking out bedrooms, even approaching me to sniff my socks. How it got in I'll never know. It disappeared momentarily, then reappeared with a mouse in its mouth, and gratefully, I opened the sliding door and let it out. It appeared completely at home. Could I have tamed it? Who knows?

I could cite innumerable examples of cultures all over the world who astonished the first explorers or colonizers with their obvious love for and attention paid to an extraordinary number of animals with which they shared their living space. Some, like the Guajá, did indeed kill and eat some of the same species they kept as pets, but what seems to have been a widespread rule is that once a pet, always a pet and never eaten. In fact, as James Serpell, an anthrozoologist at the University of Pennsylvania, has noted, "Even when well-intentioned Europeans pointed out the potential economic uses of pet animals, few of these cultures took their ideas seriously." They actually laughed at the idea.[6]

I'm tempted to conclude from the sheer volume of such observations that modern humans—if not our ancestral relatives like the Neanderthals or even *Homo erectus*—could well have taken animals in and treated them as pets, as we do or even better. The question that looms, though, is why.

– CHAPTER 4 –

Why Do We Keep Pets?

We are still searching for the answer to this simple question: Why do we keep pets? Pet-keeping turns out to be a powerful instinct, driven by cultural whims, fueled by a variety of factors, some easily noticed once you become aware of them, some so subtle as to defy discovery.

It begins with the fact that humans have always lived in an animate world, surrounded by animals and plants of all descriptions, from long, long before we were even remotely human. The life of others is the context for our evolution. We've killed, eaten, studied, and protected it. We've hidden from it and been awed by it. Today, we do all of that and also give it shelter. As a result, living things have taken up permanent residence not just in our homes but in our brains, and you don't have to look too far to find experiments that vividly demonstrate how wired for awareness of other life-forms we are. Here's a selection:

One of my favorite such studies was published in 2007 in the *Proceedings of the National Academy of Sciences*. Volunteers took part in an experiment testing a phenomenon called change blindness. They were shown a series of nearly identical slides, each followed by a brief blank slide, and were asked to identify the one element in each slide that had been removed, then restored. This sequence continued until the change was identified, then on to the next slide. Tests like this are really challenging, misled as we are by our conviction that we take in all the detail of an image at a glance—it is just not true.

In this version, both inanimate objects (silos in a field, mugs on a desk) and animate (both humans and animals) disappeared, then reappeared. The results were clear: Changes in living things were consistently detected quicker and more frequently than changes in the nonliving. However, it seemed that experience wasn't the reason: Automobiles composed some of the targets, but even though we learn about cars early in life—especially, as the authors put it, "the life-or-death importance of anticipating their momentary shifts in trajectory"—cars were definitely not spotted as quickly as animals. If not recent experience with objects (cars didn't really become a threat until the twentieth century), then what? Experience over vast stretches of time. Whether threat or attraction, animals stand out in our perception.[1]

Taking it up a notch was a 1944 experiment conducted by a pair of psychologists, Fritz Heider and Marianne Simmel. They created a simple ninety-second animation, crude by today's standards, depicting two triangles, a circle, and a large rectangle. The rectangle remains still, although a section of one side periodically opens and closes. The smaller geometric figures move around, sometimes entering the rectangle and leaving again. And that's it. Ninety seconds of moving geometric figures. But you *must* watch it to feel the effect![2]

Heider and Simmel presented the video to a psychology class and asked them simply to "write down what happened in the picture." Of the thirty-four students in the class, only one stuck to the directions and described what she saw this way: "A large solid triangle is shown entering a rectangle. It enters and comes out of this rectangle, and each time the corner and one-half of one of the sides of the rectangle form an opening. Then another, smaller triangle and the circle appear on the scene." Fair enough, but the other thirty-three students saw it quite differently. Here's one example:

Why Do We Keep Pets?

A man had planned to meet a girl and the girl comes along with another man. The first man tells the second to go; the second tells the first and he shakes his head. Then the two men have a fight, and the girl starts to go into the room to get out of the way and hesitates and finally goes in. She apparently does not want to be with the first man. The first man follows her into the room after having left the second in a rather weakened condition leaning on the wall, outside the room. The girl gets worried and races from one corner to the other in the far part of the room.

A reminder: The video includes only two triangles, a circle, and a rectangle.

Some of the students were also told to interpret the movements as the actions of persons and asked what kind of persons they were (answers that the first group had already provided unprompted), and they had no trouble characterizing the larger triangle as "aggressive, warlike, pugnacious and belligerent."

There appears to be no limit to the interpretations. The University of Southern California Institute for Creative Technologies persuaded comedians to give their own insights into the Heider-Simmel video, and here's what they came up with: "It seems like he's aggressively yelling at her. He has gross alcohol breath right now and it's about eleven a.m." (Jake Weisman). "So I think what I'm saying is that the little circle and the triangle are having an affair. He's pushing back and the circle is sneaking inside of the box" (Hunter Cope). "It's like a home invasion situation going on. The circle is trying to sneak into his house" (Paul Cibis). "There's something creepy about the way he's opening the door—the little triangle" (Hunter Cope).[3]

It is the comedian's job to take things a little further than most of us, but there is an abundance of (serious) psychological interpretations of Heider and Simmel's experiments too. We apparently have an urge to animate the inanimate. One of the key features seems to be that the objects in the video appear to be self-propelled, in contrast to, say, a leaf blown by the wind. Some argue that it means that we have an uncontrollable urge to tell a story about something that really has no story attached to it; others argue that while there's always a story to be told, the details vary dramatically from person to person, even though the shape of the narrative remains roughly the same. Heider himself suggested that we have a tendency, even when cautioned against it, to apply human qualities like thinking, emotion, and intention automatically, even to triangles and circles bumping into each other in a video.

So, one experiment illustrates our attention is drawn to other animals above all else, and another proves that we are pretty skilled at giving geometric figures human personalities and intentions. Both must play a role in the thought process underlying the desire to keep pets.

And of course, as I mentioned in chapter 1, there is biophilia, the idea that E. O. Wilson argued: "From infancy . . . we learn to distinguish life from the inanimate and move toward it like moths to a porch light."[4]

Biophilia has the effect of taking the experiments I just described and focusing them into an attraction, an attraction that, according to Wilson himself, has a close connection to pets. When a biologist for the ages like E. O. Wilson argues that biophilia (an attribute that might have some innate qualities but is also shaped by culture and experience) contributes to pet-keeping, you have a powerful basis for thinking that pet-keeping isn't just a product of the last few centuries or even millennia. It has likely played a role in our lives for an incalculable span of time, incalculable because the evidence for it will likely remain forever beyond our reach.

Our longtime close relationship with our pets may persuade us we know more about them than we really do. Anthropomorphism, or reading thoughts into an animal's mind when it might not be justified, is an example, and it has had a long up-and-down history. (Fritz Heider even attributed his students' compulsion to create a story out of the "lives" of the geometric figures to anthropomorphism, although it is obviously an extreme version of that.) Charles Darwin claimed that "lower animals, like man, manifestly feel pleasure and pain, happiness and misery," even though he went on to moderate those comments slightly by writing "man . . . is capable of incomparably greater and more rapid improvement than is any other animal . . . and this is mainly due to his power of speaking and handing down his acquired knowledge."[5] Darwin was, even in some modern views, pretty liberal, but in the early twentieth century, things turned.

In 1913, the psychologist John B. Watson called Darwin's (and others') support for comparing animal thinking to human "absurd." Watson was the founding father of behaviorism, the idea that human or animal behavior could be evaluated only on the basis of actions, not on what might be going on in their minds. But again, as attitudes evolve, the most rigid version of behaviorism has given way in the century since Watson, and today, many researchers are willing to extend consciousness or sentience to a range of animals, including birds and even invertebrates like octopuses.

Of course there are limits to our ability to be confident about what's going on in an animal's mind. The philosopher Thomas Nagel made these limits vivid with his 1974 essay in the *Philosophical Review* called "What Is It Like to Be a Bat?"[6] He argued that inferring mental activity from simple observations of behavior didn't make sense because this sort of thinking "could be ascribed to robots or automata that behaved like people though they experienced nothing"—we hear echoes of that argument today in response to claims that AI can be sentient. His argument was *not* that you could feel just like a bat if you imagine yourself being very light and furry, with relatively weak eyesight (but not, as some argue, blind), with enormously extended fingers with webbing that allow you to fly, and with incredible navigation equipment capable of analyzing echoic patterns of ultrasonic clicks and being able to detect, track, and capture fast-flying insects. He threw cold water on that approach: "In so far as I can imagine this (which is not very far), it tells me only what it would be like for me to behave as a bat behaves. But that is not the question. I want to know what it is like for a *bat* to be a bat." Nagel concludes we're just not equipped to do that, something you should remember when you think your dog feels guilty or your cat is resentful.

Even though Nagel drew a line in the sand when it comes to imagining you can think or feel *exactly* like another animal (you can't even think exactly like another human!), that hasn't prevented the gradual encroachment on that line. Some of the most recent science, while not dealing specifically with pets, has shown just how dramatically we used to underestimate the intelligence of other species. For instance, a bumblebee's brain is about the size of a large grain of sand. Nothing much going on in there! Yet recent experiments revealed that a bumblebee can teach another bumblebee the solution to a puzzle. Here's how it worked: Bumblebees were taught to press two different levers in succession to get at a sugar reward, but they were able to master the two steps only if each provided a reward. If the first step didn't, they'd stop. However, if both steps were rewarded with sugar, some bees learned they had to press both levers in succession. These were now "demonstrator" bees. In the absence of the double reward, untrained bees couldn't solve the two-step puzzle, but if they observed a demonstrator doing it, some of them could. The researchers recognized this as "social learning": The bees could learn from others, even though they couldn't learn on their own.

It's studies like this that prompted researchers attending a recent meeting called the Emerging Science of Animal Consciousness at New York University to declare that "evidence indicates at least a realistic possibility of conscious experience in all vertebrates (including all reptiles, amphibians and fishes) and many invertebrates (including, at minimum, cephalopod mollusks, decapod crustaceans and insects)." The declaration isn't arguing that all these creatures have minds like ours—that really would be unbelievable—but they could likely experience pain, hunger, pleasure, even amusement.

Why Do We Keep Pets?

It's a dramatic turn: Over the years, scientists have been very reluctant to allow that animals might share some of the mental capacity of humans, but this declaration shows just how much they're coming around to that belief. Of course pet owners beat them to it; they've likely been thinking that way since the first animal became a pet. It's an important part of pet owning, sharing space and time with a living thing with which you think you might have something in common. (As an anthropomorphic coda to my visiting-weasel story from the last chapter, I invited the visitor as it went through the door to come back any time, especially if it were bent on catching another mouse.) Prior to solid scientific evidence that there's actually something to this, pet owners' thinking like this was dismissed as mere anthropomorphism, as wishful thinking. Now we're more open to the reality of animal sentience, but that still leaves open the question of why pets? It has always taken a lot of effort to keep one. What's in it for us?

– CHAPTER 5 –

Does Pet-Keeping Even Make Sense?

The legendary twentieth-century biologist J. B. S. Haldane is usually credited with saying, "I would gladly give up my life for two brothers or eight cousins" (although there is some uncertainty about the origin of the quote). Nonetheless, it leads us to a fascinating, if somewhat foreign, way of thinking about pets. The quote is a shorthand way of explaining how evolution works. Yes, it *is* all about passing your genes on to the next generation (the goal for every living thing on earth except those humans who exert choice), but those genes don't have to flow through you—it's a little subtler than that. Because you share 50 percent of your genes with your siblings, and 12.5 percent with your cousins, Haldane's quote applies: If you fail to reproduce because you've given up your life, but all eight cousins have their own offspring, you may have sacrificed your own offspring, but you're still represented in the genetic contribution to the next generation.

What does evolution have to do with pets? Or, rather, with pet-keeping? Pet-keeping today is a universal human activity, and while it's impossible to know how long-ago humans—or even our nonhuman ancestors—began keeping pets, you can be sure it was at least a few tens of thousands of years in the past. We'll probably never know exactly, because the materials that could provide evidence, like bones

Does Pet-Keeping Even Make Sense?

and burials, are simply harder to find as we probe further into the past, when human (or humanlike) populations were much smaller, and earthquakes, floods, and volcanoes have obscured what might once have been visible. But enough time has surely passed for humans to have evolved—could pets have had some influence on that process?

In the last chapter, we saw that we are sharper at detecting animals than practically anything else, an ancient, innate feature of the human brain. It's easy to argue that being sensitive to sudden movements of animals—prey or predator—would have been of survival value to hunter-gatherers. Are there other brain mechanisms related to pets that might have been retained because they bestowed survival advantages? What about the idea that we tend to anthropomorphize, whether it's turning a video of geometric objects into some sort of human-centered rom-com or, even after Thomas Nagel shows you just can't, believing that we can imagine the lives of other animals? The latter is definitely part of having a relationship with a pet, especially one that feels closely related, such as a dog or cat. We think we know what they're thinking or feeling. My knowing what you might be thinking or about to do is an important social skill. And there's no reason to think it wouldn't be transferable to our relationships with other animals.

Steven Mithen, an archaeologist and professor of early prehistory at the University of Reading, uses that idea to suggest that the ability of early humans to imagine the thought processes of animals was an important factor in the survival of those humans. This ability would, of course, apply equally to predators and prey and would be complex, including everything from when the wolf pack was likely to come by, to where the birds were nesting, to what the signs were that a cave bear was going to attack.

This would have been an early form of anthropomorphism, and Professor Mithen contends it differentiated early modern humans (people like us) from the Neanderthals. Mithen suspects that Neanderthals probably had the ability to know what *each other* was thinking but weren't able to transfer that skill to thinking like an animal. That could have contributed to their extinction. On the other hand, if this sort of anthropomorphism was a useful addition to the human brain, it wouldn't be surprising to see that we still have it—and that it plays a role in our relationships with pets. Common knowledge suggests it's rampant.

However, there's another intriguing detail to the idea of attributing humanness to animals beyond simply inventing their thoughts for them. We do that consciously, but we also fall into the unconscious trap of being attracted to their humanlike qualities, especially when they are young—puppies, kittens, piglets, ducklings. It's called cuteness, and we can thank the legendary Konrad Lorenz for this idea.

Lorenz, an Austrian Nobel Prize winner, is most famous for his work on innate behaviors and the idea of a critical period—imprinting—when such behaviors are established for life. You often see photos of him being followed by ducks who shortly after birth became imprinted on him rather than their mother and never really shook that idea.

Relevant to pets, Lorenz established the idea of cuteness: that is, that baby animals, including baby humans, have a unique facial shape and structure that attracts and encourages maternal behavior. These features include a high forehead, large round eyes, and chubby cheeks. Humans, seeing a human baby (who, of course, has all of these), are immediately and apparently automatically and helplessly triggered to say "aww" and slip into baby talk.

There is abundant scientific support for this idea of cuteness. One of the best studies had undergraduate students (the experimental animals of psychology) look at baby pictures and rate them for cuteness and for their own desire to take care of them. The participants, unbeknownst to them, were rating three different versions of each baby's face: unretouched, altered to be "less cute," and altered to be "cuter." As the researchers suspected, the big-eyed, chubby babies elicited more ratings of cuteness and more expressions of desire to care for them,

Does Pet-Keeping Even Make Sense?

although there was one interesting discrepancy: There were no sex differences in cuteness ratings, but female undergrads expressed stronger desires to care for the cute babies. This has been seen as indicative of an evolutionary pull on females as principal caregivers, whether as mothers, grandmothers, or even female siblings.[1]

A slighter more off-the-wall example of the attraction of baby faces for both human sexes comes from the world of entertainment.

Teddy bears (the actual toys, not just the picture-book versions) first appeared in the United States in the early 1900s. The earliest versions were much like adult bears, with small eyes, low foreheads, and long snouts. But these bears evolved very quickly—over the next twenty years, they morphed into cub-like teddies, with the Lorenzian "cuteness" factors built in. They became a staple of the toy-buying world.

The evolutionary scientist Stephen Jay Gould pointed out that Disney's Mickey Mouse was subjected to the same radical facelift, beginning his career in the late 1920s as a mouse who looked more like a rat, but as the years passed, his eyes widened (actually, the original eye became merely the pupil of the updated one), his snout thickened to make it look shorter, and his eyes moved up, giving his forehead a more rounded appearance—a significant feat because Mickey's head had been circular from the beginning.

Both the generic teddy bear and Mickey Mouse changed facially to increase their cuteness and therefore their appeal. I doubt that the teddy bear makers or Disney animators knew that they were duplicating the evolutionary process called neoteny, a set of changes through time that ensure that the adults of today resemble the juveniles of the past. But I'm sure they knew they were making their animals more charming, if only because consumers told them so.[2]

Putting together the evidence from the entertainment industry and the psychology lab establishes that humans prefer baby-like features and, more than that, may feel more nurturing to human babies whose features exaggerate those qualities. It makes total evolutionary sense for the facial features of newborns to encourage their mothers—and even their fathers—to be attracted to them and to care for them. That is a straightforward survival mechanism, making it more likely that the parental genes will survive in the offspring. There's really not much debate about that. But I think you can see that a logical offshoot of this is that cute baby *animals* might just turn on the same responses in humans that the humans' own babies do.

Do they? From the strict evolutionary point of view, it doesn't appear to make a lot of sense. We nurture our babies because they carry our genes (see Haldane's quote at the beginning of the chapter), but dogs and cats have nothing to do with our genetics. Responding to cute features in puppies and kittens, again through the lens of evolution, seems to be a waste of time. Any practice that didn't benefit human survival is unlikely to have lasted through the constant evolutionary sifting of survival positive from survival negative. Putting emotional and material resources into raising animals of another species is puzzling, especially at a time tens of thousands of years ago when life was likely more precarious than today. Or does it?

Put it in the context of our ancestors living a nomadic hunter-gatherer life, where it's very important for a new mother to lavish attention on her newborn, and to have that attention turned on by the sweet baby face of a child is a very good thing, a survival technique, an innate mechanism that betters the chances for that baby. At the same time, because the group is small, there aren't hordes of other, unrelated children in the group, so the attentional mechanism doesn't have to be precise and fine-tuned. The mother's attention is unlikely to be distracted by the very few other cute faces.

However, when families of wild animals are captured and brought into contact with humans—a widespread activity around the world, past and present—and there are young animals among them with baby-like features, this attraction is easily co-opted. Because the

Does Pet-Keeping Even Make Sense?

response to cuteness has never needed to be so finely attuned as to exclude the babies of other humans, let alone animals, suddenly puppies, kittens, et al., are welcomed.

There is some evidence that attention to babies versus attention to pets are in conflict: One study showed that women who were primed to pay attention to their own offspring (they were either pregnant or had given birth) felt less attached to their cat.[3] I'm not aware of any other study that actually measures how attachment to a pet might wax or wane in situations like that.

Put this way, it looks like the animals we keep as pets—as least the ones who, when young, have faces that turn on this human attraction to them—have lucked into an innate human mechanism that increases their chances of survival. (It shouldn't be necessary to point out that some pets, like goldfish, turtles, and lizards, don't benefit in the same way.)*

If these mammalian pets have piggybacked onto a human nurturing mechanism to their advantage (and the countless observations of women around the world nursing baby animals testify to this[4]), that suggests that their evolution into pets has been a good thing—for them. And if you pursue this evolutionary way of thinking, you could conclude that it might not be advantageous for us.

Some scientists have enthusiastically pursued this argument of "humans as suckers" by comparing the pet situation to wildlife that practices what's called brood parasitism. Many species of birds, fish, and insects enhance their survival by laying their eggs in the nests of other species, thus saving the effort of building and defending a nest against predators and feeding and raising nestlings. When it comes to birds,

---

* I don't count the innumerable posts on social media exclaiming how cute lizards are. If you think your lizard is cute, I have no quarrel with that, but for me, the threshold is, is there science that supports this?—as there is with puppies, kittens, baby rabbits, etc.

in North America, the brown-headed cowbird is the most common nest parasite; in Europe, it is the common cuckoo. Although there is a wide variety of behavior involved in nest parasitism, generally a female chooses the nest of a target species and lays one or more eggs in that nest. The female host bird may or may not reject the foreign eggs and destroy them, or may even build a whole new nest on top of the original, but often none of those defensive actions is taken and the host female raises the intruder's offspring as her own. Amazingly, at least to the human eye, she fails to recognize what should be some obvious warning signals: The uninvited egg often looks different from her own, and the nestling, once hatched, usually looks dramatically different from the host female's nestlings, sometimes being double *her* size. Nonetheless, she faithfully feeds the foreign bird until it reaches adulthood and leaves the nest.

Most often, the host bird is directly disadvantaged, either by the foreign nestling hogging the food the female brings back to the nest for her own chicks, or outright killing them. Either way, she loses the opportunity to introduce her genes to a new generation. To ensure raising a complete brood of her own, she might have to start over, wasting valuable nesting time. It's a serious evolutionary disadvantage to the host bird, while of great benefit to the parasite. In cases where the host bird mounts no effective defense, it is losing what's called the coevolutionary arms race.

The similarity to pet-keeping relates to the human's caring for the baby animal based on the innate preference for "cute" features. It's the cowbird; we're the host. It's an automatic response, and while not intended for other species (or really even the children of other humans), it's not refined enough to prevent that. In the bird example, the host female can be misled by the parasite's nestling having similar plumage or mouth markings evident when begging for food or even calls for food.

Does Pet-Keeping Even Make Sense?

For many host birds, fighting back is complicated, labor intensive, and, in the long run, just not worth it. In the human-pet situation, the equivalent action (if indeed it's an evolutionary disadvantage to adopt a pet) would be to override the natural attraction to cute faces. But that could have a direct negative impact on parenting, which would be an evolutionary misstep. It must be that the value of the pets to humans outweighs the disadvantages to humans spending parental time on pets.

Painting humans as evolutionary victims of pets might be a stretch, and I should add that one of the acknowledged experts on pet issues, James Serpell of the University of Pennsylvania, has pointed out a straightforward objection to the argument. Noting that birds that raise the nestlings of cuckoos and cowbirds are victimized because they are *incapable* of differentiating them from their own, he pointed out that "Pekingese and pugdogs may be cute but it is ludicrous to suggest that people are seriously incapable of distinguishing between them and real babies."[5]

All of the above ignores the issue of attention to the pet after it has become an adult and looks like a dog, not a baby! But by the time that happens, our ability to anthropomorphize has taken over: We think we know what they're thinking—we infer it from their behavior—and that allows us to fill in the gaps left by the pet's inability to speak. And make friends with them. In this way, if indeed pets took advantage of our nurturing instinct when they were babies, they continue to benefit because of our insistent belief that we know what's going on in their minds.

But are they really taking advantage of us? Or are the benefits to us equal to (or even greater than) the benefits we supply?

Before we tackle that thorny question, I leave these final words to psychologist John Archer from his paper "Why Do People Love Their Pets?"

The following question needs to be answered in the future: whether pet-keeping is like religion, where fitness costs are outweighed by long-term positive influences on self-esteem and self-confidence; or whether it is like drug-taking, where the long-term fitness costs outweigh any short-term beneficial mood changes, yet once established, its cessation produces negative feelings.[6]

– PART II –

The Unlikely Victors

– CHAPTER 6 –

Where Did Dogs Come From?

Whether you're a dog-lover or not, at the very least they deserve your respect. This is, after all, the *first* animal ever to be domesticated. A long list followed—cats, sheep, goats, cattle, pigs, horses, rabbits, chickens, ducks, and camels—but the dog beat them all by thousands of years. Many have made the comment that in the absence of domestication, the world—our world—would look very different. And while most of those other animals (aside from the cat) were tamed primarily to serve a useful purpose, not to be pets, dogs are now both pets and service animals.

Modern dogs are descended from wolves, but exactly how and when that happened is still contentious. And mysterious. I mean, how on earth can you go back thousands of years and deduce how a wild animal gradually became domestic? There are two primary kinds of evidence, genes and fossil bones, and plenty of speculation.

Fossil bones are a case in point. No investigation of the distant past using them is without uncertainty, but the origin of the dog is even cloudier than human evolution, where the skeletons of our ancestors/relatives, even the most recent, the Neanderthals, can be distinguished pretty clearly from physically modern humans. For instance, Neanderthal skulls sport a prominent brow ridge over the eyes, a face that protrudes forward, and a bulge at the back of the skull, none of which is obvious in modern humans. One of these is often enough for a positive identification, but many excavations of Neanderthal campsites also contain tools that are stylistically different from those made by humans who might have been living at the same time and in the same area.

By contrast, the differences between wolf skeletons and the bones even of modern dogs, especially huskies, malamutes, and German shepherds, are not so clearly different. The discrepancies are subtle to the eye and require the application of precise calipers and subsequent comparison with dozens of other measurements. If you just glanced at the two skulls side by side, you could likely make the observation that the wolf skull looks a bit stretched, the slope of the muzzle in front of the eyes more gradual, the teeth spaced slightly farther apart. But casual glances fall far short of what's necessary to be able to come to actual conclusions.

For example, one published set of measurements of details like the width and length of the jawbone and snout, differences in the size of individual teeth, and a further set of what the researchers called "micro-anatomical differences" revealed that only a handful of such measurements could reliably distinguish a dog's skull from a wolf's. On those grounds, many ancient remains previously identified as early dogs were likely wolves instead. And of course there were inevitably animals whose bones were preserved when they were only partway through the drawn-out wolf-to-dog transition.

Amazing advances in the ability to pull DNA samples from fossil bones have established genetics as a powerful alternative for pinpointing when wolves and dogs went their separate ways, but here too there are limitations. Modern wolves and dogs share 99.9 percent of their genes, so even though ancient dogs and wolves are bound to be different, it's likely that those differences are still extremely slight. Upon these tiny differences rest significant debates.

Where Did Dogs Come From?

Now that we're armed with bones and genes, what can be said about how ancient wolves transformed into today's dogs? One thing is certain: The transition was anything but an overnight sensation. Instead, it took millennia, a long period of time complicated by gaps in the fossil record, clouded by the inevitable hybrid offspring created when early dogs mated with wolves, and further messed up by the fact that the wolf thought to have been the immediate dog ancestor is now extinct.

That ancestral wolf lived in northern climes—as modern wolves do today—and while it's still impossible to pinpoint exactly when the transformation began, a date of about 30,000 to 35,000 years ago might be about right. At that time, the last traces of the Neanderthal people (if they were still around at all) were eking out a living on a European landscape that might have been a lot more hospitable than you might think. Yes, it was the height of the last major ice age, but there wasn't ice everywhere, and the remains of plants found in the stomachs of mammals like mammoths, woolly rhinos, and horses suggest that even in winter, a sunny day in the Ice Age might have been like a modern—and comfortable—alpine meadow.

Modern people, pretty much like us, had recently arrived in Europe and were on the rise. Could either Neanderthals or modern people have been the first dog owners?

Given both the genetic and bony resemblances between wolves and dogs, pinning down a date of origin is extremely difficult, but it's about that time when doglike wolves, or wolflike dogs, begin to turn up in archaeological sites.*

As you might expect, these earliest hints of something dramatic happening are so slight that conclusions are very hard to draw, but

---

* Some researchers have been tempted to call these animals wolf-dogs, but today that term is commonly applied to modern hybrids of contemporary dogs and wolves—not at all the same thing.

besides skeletal signs, like shortened muzzles and a slight crowding of teeth as a result, there are other clues. For instance, microwear on the teeth indicates that these doggy animals were chewing on bones more than their wolf contemporaries were. Also, these new animals are estimated to have been roughly four or five kilograms (ten pounds) lighter than the wolves of the day.

It's important to remember that the shortage of definitive remains, spread over many thousands of square kilometers (millions of acres) of territory across northern Europe and Russia and tens of thousands of years, means that the where and when of this process is sketchy, to say the least, accounting for the uncertainty and debate. After all, what looks like an animal on its way to doghood may be just a random and short-lived variety of wolf. For a sense of what the search for a definitive story is like, here's a short passage from the Wikipedia page on "Paleolithic dogs" detailing the progress of knowledge:

> A first article proposing the Paleolithic dog, its refutation, a counter to the refutation, a second article, its refutation, a third article that includes a counter to the refutation, its refutation, a counter to the refutation, another refutation . . . "[1]

Take my word for it: The research is complex and fascinating (if sometimes bewildering), involving not just the wolves thought to be the ancestors of dogs and those (possible) first dogs, but all the other animals common at the time, from mammoths to cave bears and humans. Fascinating but dense.

As thousands of years in the archaeological record pass, the fog clears somewhat with the Bonn-Oberkassel canine, the oldest unambiguous dog ever found. Its skeleton has been dated at 14,223 years old, with an uncertainty of about fifty years in either direction. Obviously, the oldest dog ever found is fascinating enough (until an older one is found), but the Bonn-Oberkassel dog stands out for another reason. It was buried together with the remains of two humans (owners?), but it was quite a young dog—less than six months old—and showed distinct signs of canine distemper, a disease that likely killed it. There appears to have been more than one flare-up of the disease during the dog's life, any one of which should have killed the pup even earlier than it did. It seems that it had been kept alive, probably by the humans with whom it was buried, suggesting this dog was a pet in the true sense of the word. This is really a crucial find: humans caring for an animal many millennia ago.

Even though the Bonn-Oberkassel dog is unlikely to be the oldest we will ever find, for the moment, it is acknowledged to mark the end to the process of domestication of the wolf and the true beginning of dogs—but far from an absolute end. Intermixing of wolves with dogs didn't stop then and in fact continues today where the two species encounter each other.*

How Did This Happen?

The evolution of dogs from wolves is an utter transformation of a fierce, independent predator to a human companion. So dramatic that it boggles the mind.

There are two competing theories to explain the transformation. Both, even if speculative, must account for the apparent timing and location of the event. More details to come, but the key difference between the two is that in one, wolves take the initiative and accommodate to humans, while in the other, humans take the first step by

---

* I'm hedging about Bonn-Oberkassel being "the oldest dog" partly because archaeology never ends, and partly because the remains of a 13,000-year-old dog have been found in a cave in Haida Gwaii off the north coast of British Columbia. One thousand years isn't a lot of time for the earliest dogs to move from Europe to North America, which suggests the possibility that they had been around for much longer.

adopting wolf puppies. Essential to both is that the wolf-to-dog transition means that wild wolves become, if not pets, at least comfortable in the presence of humans. Ultimately, that requires underlying genetic changes.

The involvement of genes in this transition, although previously suspected, was established by a famous set of experiments on foxes conceived and executed by Russian scientist Dimitri Belyaev, first in the Soviet Union and later in Russia. His fox study began in 1959, and while Belyaev died in 1985, the experiments continue to this day. Belyaev worked with undomesticated silver foxes, but controlled their breeding in one straightforward way: In each generation, he allowed only the tamest animals to have litters. That is, only the 10 percent that were the most comfortable with humans—either allowing themselves to be approached or actually doing the approaching—were permitted to have offspring.

Belyaev suspected that by doing so, those foxes that were genetically predisposed to being tolerant of humans would pass that trait on. That is exactly what happened. After only ten generations (one litter per year), Belyaev had succeeded in breeding a whole new kind of silver fox, one that behaved like a dog, approaching people it liked, licking their hands and faces. But that wasn't all: In a somewhat unexpected turn (although Belyaev had wondered if this might happen), the foxes began to change physically as well. Upright ears became floppy, coat colors changed, and tails became more upright—all features you see in any off-leash park today. It seemed like wild foxes were not only well on the way to behaving like dogs; they were starting to *look* like dogs too.

The Belyaev experiments have been criticized by scientists who point out that he acquired his original foxes from a fox farm in Prince Edward Island. Photographs and accounts of the behavior of those foxes while still on the farm suggest they weren't *completely* wild at all. However, the dissenting researchers' argument isn't with the idea that physical changes accompanied behavioral changes, nor that selective breeding can solidify the trait of tameness. Instead, they contend that the set of changes is not necessarily true for *all* domesticated species. The core of the Belyaev experiments still holds: that there are, among any population of untamed animals, some that are innately more willing to engage with humans, and that with the right breeding, such tameness could become firmly established.

The Belyaev experiments, while important, don't really do much to differentiate between the two competing ideas of the creation of the first dogs. The idea that tameness can be established by selective breeding and continued association with humans is consistent with both theories: That hunter-gatherers collected newborn wolves from dens or that wolves gradually adapted themselves to being camp followers.

There is a superabundance of scientific literature tugging this way and that around these two alternatives, but here's a quick sketch of both scenarios and the evidence supporting or casting doubt on each.

One idea is that human hunter-gatherer groups (this was all happening long before our ancestors settled down and began to farm) left behind middens—heaps of meat scraps, bones, and trash of all kinds—as they moved across the landscape in pursuit of game (we've always been good at that). Packs of wolves, mobile as they are, were attracted by this novel source of food and began to follow and even hang around the hunter-gatherers, cautious, to be sure, but bold enough to dart in and grab food whenever the coast seemed clear. With time, both species would grow accustomed to being in the company of the other. The benefits at first seem to be largely on the wolves' side, but it might have been true that the people learned that having wolf packs close by would alert them to the presence of other predators or prey. Those wolves most comfortable being at close quarters with humans might pass on genes for that tolerance, and slowly, over a few wolf generations, an attachment to humans could develop.

There are some serious doubts that this could be the origin of dogs. For one thing, such wolves hanging around humans are not domesticated—they're at best tolerant or habituated. How that would progress to something more intimate isn't clear, given that wise humans would have been extremely cautious around wolves that had overcome their fear of humans but were still semi-wild.

There's also a practical issue: Wolves need a lot of protein every day to maximize reproduction, and the waste left behind by a thinly scattered human population might not suffice. But here, as always, there's a counterargument. Humans can tolerate only about 20 percent of their diet as protein, depending on fat and vegetable matter for the rest. So a large carcass, like a mammoth, would provide way too much protein for a small hunter-gatherer band and might just make up the protein needs of the wolf followers.

The competing theory is radically different. It envisions the same hunter-gatherers gathering up very young wolf pups from dens and raising them as companions, pets, and even hunting collaborators.

At first glance, it seems weird that ancient humans could get away with stealing wolf cubs from their dens, or even be tempted to do so. But as I mentioned in chapter 3, a variety of hunter-gatherer people from around the world are known to have adopted the young of a variety of species—it is not an uncommon practice. And there's evidence that the same could have been done with ancient wolves.

Of all animals, why domesticate wolves, one of the fiercest of them all? It could be a mistake to assume that the wolves of today, having endured persecution from humans for centuries, represent the wolves of thousands of years ago, when there were so few humans on the ground.

Naturalist Francis Galton, Charles Darwin's second cousin, quoted the explorer Samuel Hearne's eyewitness reports of First Nation peoples' interactions with wolves and their litters:

> They always burrow underground to bring forth their young, and though it is natural to suppose them very fierce at those times, yet I have frequently seen the Indians [sic] go to their

dens, and take out the young ones and play with them. I never knew a Northern Indian [*sic*] hurt one of them; on the contrary, they always put them carefully into the den again; and I have sometimes seen them paint the faces of the young wolves with vermilion or red ochre.[2]

Scientists who have spent time on Ellesmere Island in the Canadian Arctic have observed that wolves there are quite comfortable being at close quarters with them. David Mech of the US Geological Survey spent more than two decades on Ellesmere and recorded numerous occasions when wolves were completely tolerant of his presence. Mech attributes this to the fact that these particular wolves had had few, if any, encounters with humans before, and so hadn't developed the fear of humans that wolves almost everywhere else in the world have. (Although it must be said that being tolerant of humans is very different from modern dogs' predilection to being attracted to them.)

Mech argues that even fearful wolves, if captured as cubs and then raised, fed, and protected by humans, could become tamed, especially if only the most social were allowed to breed. And surprisingly, wolves don't aggressively defend their dens from humans. Mech and his colleagues had no trouble approaching pups when the parents were off hunting.

For either of these scenarios to provide a route to domestication, they must somehow include a mechanism for controlling breeding, à la Belyaev. Pups collected at a very young age from a den are inevitably going to be kept close by their humans and fed by them. This would likely reduce both the opportunity and tendency to leave the group and mate with wilder wolves. After a few generations, wolves like this accommodate to living with humans. If for any reason individual wolves failed to become tame, they would likely be driven away or even killed.

Funnily enough, the argument that humans don't need as much

protein as wolves fits this scenario too—this difference minimizes competition for food and makes coexistence much easier to manage. While this scenario of active adoption seems to me to be a better fit with what actually seems to have happened—that over time, wolves recruited as puppies gave rise to dogs—there are questions. One of the tricky aspects is that for wolves to adjust as fully as possible to living with humans, they must begin to be socialized before they're weaned, around three weeks of age or so. Even if that's accomplished, modern attempts to do so reveal that the adult animals still have wolflike traits.

Many millennia were to pass before these early dogs became the modern breeds that we're familiar with. There are a handful of modern breeds that are genetically closer to their wolf ancestors than the numerous others recognized by kennel clubs, and while some of those make intuitive sense, like Siberian huskies and Alaskan malamutes, some of the others are as unwolflike as you could imagine: shih tzus, Pekingese, basenjis, and chow chows. These breeds can be as much as several thousand years old (basenjis appear in paintings from ancient Egypt), but the majority of modern breeds were created in the nineteenth century—classic examples of human intervention, as we'll see.

A final note: You might think that modern wolf-dogs—the offspring of a wolf and a dog—would provide insights into the evolution of the dog, being as such animals are part wolf and part dog. But they can't, for a number of reasons. First, as I mentioned earlier, the ancestor of the dog was a wolf that is now extinct, so the genes of a modern wolf are not the same; maybe close to that mystery wolf, but not the same. Second, mating two modern animals is in no way a replication of the long, slow, meandering process of wolves turning into dogs tens of thousands of years ago. Not only that, the environment of thirty thousand years ago and the environment of today are radically different, and genes always respond to the environment. Wolf-dogs are intriguing, but not because they shed light on the domestication of dogs.

– CHAPTER 7 –

Where Did Cats Come From?

The history of the domestic cat is very different from that of the dog. Details aren't clear, but outlines are. From the beginning, dogs were useful in guarding and hunting, even as their humans were still nomadic hunter-gatherers. Herding was added to those roles when humans settled down to begin farming, and along the way, and crucial to the success of these early experiments, humans and dogs established a relationship. The history of the dog since those early days thousands of years ago has unwound along these lines, with workers and companions still a good rough outline of the categories of dogs that exist in human company. (It's always important to remember that the majority of dogs in the world do *not* live with humans.)

Cats are a radically different story. They too worked their way into cohabitation with humans thousands of years ago, but always with cat-like reserve. Yes, there are hundreds of thousands of house cats today that never venture outdoors, but even among them, there's a sense that if the door was left open, they might bolt. Outdoor cats, by contrast, always seem to be at risk of disappearing, perhaps even for days, their fidelity to their humans tempered by their innate nature, which surely can be traced back to their first introduction to human life.

That was probably about 11,000 years ago, somewhere in the Middle East. Our hunter-gatherer ancestors had begun to abandon the nomadic life to create permanent spaces where they could grow crops like barley and wheat. One archaeological excavation in Jordan revealed that the farmers of the time were already plagued by mice: Their grain was stored on a raised platform, suggesting they hoped that doing so would ensure its inaccessibility to mice—although my limited experience with mice suggests that would be unjustifiably optimistic.

The switch to agriculture was likely the most dramatic move ever made by humans before or since, dramatic enough to immediately alter the local environment. Whenever that happens, there are winners and losers. In this case, there was a cascade of winners: humans, because farming reduced (somewhat) the risky adventure of pursuing a food supply, and rodents, particularly mice and rats, which found stores of grain a novel and irresistible source of food.

But it's also true in nature that a sudden windfall for one or two species usually draws some party crashers, and in this case, it was cats. Crowds of mice are just as inviting as the piles of grain that attracted them. The wildcats of the Middle East, *Felis silvestris lybica*, began hanging around these early farms and becoming tolerant to living in close quarters with people. Unlike early dogs, these cats weren't working with people so much as working *near* them, but inevitably, a kind of rapprochement had to be established. Humans would have immediately recognized the value of cats, and cats had to become tolerant of humans to gain access to mice.

As with the domestication of dogs, the exact dates of when cats and humans began to tolerate each other are murky, given both the rarity of fossil remains and the near anatomical identity of wildcats and house cats. The most dramatic finding is a grave of a human and cat on the island of Cyprus, dating back 9,500 years ago (note this is at least five thousand years later than the generally accepted "first dog"). The significance of the find is that wildcats had never existed on Cyprus, so this cat must have been brought from the mainland by ship, and its burial together with a human surely identified the two as companions. (It could possibly have been a stowaway, although

Where Did Cats Come From?

given the diminutive boats of the time, it would have been difficult to overlook an onboard cat.) On the other hand, mice could have stowed away successfully, and were already abundant on Cyprus, which just underlines that the best explanation for the cat is that it was brought over on purpose, either because it had a relationship with the buried human or because it was a wildcat imported to combat the already established vermin. No other explanation makes as much sense.

It was the beginning of a beautiful, long-lasting, and more or less steadfast relationship. The cat not only has maintained its role as mouse catcher but also, of course, has expanded that to house pet, but its popularity has endured some extreme peaks and valleys.

The summit was reached in ancient Egypt just a few thousand years after these first engagements between cats and people. The Egyptians were fascinated by animals of all kinds. The cemetery at Hierakonpolis, which welcomed the burial of people and animals for centuries, beginning about six thousand years ago, contains a riotous collection of tamed and untamed animals like baboons, elephants, hippos, crocodiles, ostriches, and even aurochs (the ancestor of one of the two main kinds of cattle today). And cats. One cat in particular hints that by this time in ancient Egypt cats were being kept by humans. What's left of this cat suggests it was the same species that was originally tamed in the eastern Mediterranean (as are all living house cats). More important, it appeared to have suffered a couple of fractured bones at some point, but the bones had healed. Had the cat been living in the wild, it could never have survived long enough for that to have happened, so it must have enjoyed human care for the last weeks and months of its life. This doesn't mean this cat was anything special—Egyptologists think it was buried not to sanctify it, but to somehow benefit the interred humans nearby. Nonetheless, it was important enough to merit burial, and this was only the beginning.

Remember that what we call ancient Egypt lasted thousands of years, from five thousand years ago to the time of the Roman Empire, so there was plenty of time for cats to ingratiate themselves with Egyptians. But time enough to rise to godlike status? The two most prominent catlike gods (though there were others) were Bastet and Sekhmet, both of whom were daughters of the sun god Ra and protected him. Bastet even had the city of Bubastis dedicated to her.

These catlike gods appeared in various forms, sometimes straightforwardly as cats accompanied by a dedication to the god, sometimes as a human body with a cat's or lion's head. Why cats? Some Egyptologists argue that cats were admired for their dual nature: fierce hunters capable of killing venomous snakes but also dedicated mothers. The sun connection has also been interpreted as this dual nature from a different angle: life-giving warmth but also the dangerous, unforgiving desert sun. Cat-lovers today, I'm sure, recognize the two sides.

Art and inscriptions revealing cats as having the status of deities aren't even the most intriguing evidence—that title belongs to the cat mummies. We have no idea how many cats were mummified in ancient Egypt, but it must have been in the millions. First, it's certain that not all repositories of mummies have been found, and even some of those that were have been lost again, swallowed by earthquakes and shifting sand. And most were never excavated and analyzed carefully by Egyptologists. All that is made disturbingly clear in this 1889 article from *The English Illustrated Magazine* describing a farmer's discovery as he dug just below the surface: cats. As described by Professor W. M. Conway:

> Not one or two here and there, but dozens, hundreds, hundreds of thousands, a layer of them, a stratum thicker than most coal seams, ten to twenty cats deep, mummy squeezed against mummy tight as herrings in a barrel. . . . The surface sand was stripped off and the cats were laid bare. All sorts and conditions of them then appeared—the commoner sort caked together in black lumps, out of which here a grinning face, there a furry

Where Did Cats Come From?

paw, there a backbone or row of ribs of some ancient puss, stood prominently forth. The better cats and kittens emerged in astonishing numbers, and with all their wrappings as fresh as if they had been put into the ground a week, and not thirty centuries before. . . . The path became strewn with mummy cloth and bits of cats' skulls and bones and fur in horrid profusion, and the wind blew the fragments about and carried the stink afar.[1]

If all that weren't bad enough—and it was—the remains of an estimated 180,000 cats, nearly 18 tonnes' (19 tons') worth, were shipped to England to be used as fertilizer.[2]

This extraordinary example is one of many, and repositories of cat mummies are spread over millennia. For instance, a pet cemetery from a much later time when Rome had already superseded Egypt turned up 536 cats, 32 dogs, 15 monkeys, a fox, and a falcon.[*]

Many analysts have described the Egyptian fascination with cats as a cult. It's true that cats held a public fascination, reinforced by their association with gods, but that translated into killing them, mummifying them, and then selling the mummies to anyone who wanted to make an offering to Bastet, Sekhmet, or any of their minor-league versions. Sometimes the mummy wrapping concealed nothing more than some bones, or gravel or dirt, packaged to look, at least in outline, like a cat.

On the other hand, some were practically works of art, like the mummy examined by Australian researchers using both neutron and

---

[*] A note about dog mummies: While canines were no match for the Egyptians' fascination with cats, they nonetheless were popular among mummy makers. The spectacular dog cemetery at Saqqara contains something like eight million dog mummies. Dogs, foxes, and jackals, all canids, were associated with the jackal-headed god Anubis, to whom a temple at Saqqara is dedicated. Some of the dogs were newborn or only days old, suggesting to archaeologists that the Egyptians must have had the ancient equivalent of puppy mills to satisfy the need for mummies to dedicate to the god.

X-ray imaging. The mummy inside was indeed a cat, although only the leg bones and tail had been wrapped. They were the remains of an eleven-month-old cat, and the assembly of the package had been very carefully done. Two layers of fabric, a fine inner layer and a coarse outer layer, were wrapped around the bones in a crisscross pattern. At the top of the mummy, the fabric had been wadded together to create a "head" complete with ears, and the entire mummy had been dabbed with red and green paint. It's hard to imagine such care would have been devoted to a mummy put together to sell as quickly as possible to tourists shopping at the annual festivals, whose raison d'être was to dedicate thousands of mummies to the gods.

It's hard to wrap (sorry!) your mind around the apparent contradiction of cats being elevated to godlike status only to be raised in kitten farms and slaughtered, just as it's difficult to reconcile painted images of cats seated under their mistress's chair dining on duck or sunning themselves in a window with excavations uncovering thousands of other members of their species, whose necks were broken to create more mummies. But we can't put ourselves in the minds of the people of ancient Egypt; what seems like unbearable cognitive dissonance to us made good sense to them. Besides, extreme contradiction also describes the state of cats of the Middle Ages in Europe.

Imagine this: It's Wednesday in late March in the Belgian city of Ypres. The year is 1817, and typical for Ypres, the weather is breezy and chilly, with temperatures around 5 degrees Celsius (40 degrees Fahrenheit). Famine is raging through Europe, but even so, a celebratory crowd has gathered in the square in front of the beautiful Cloth Hall, one of the most striking, and largest, Gothic buildings of the time. The hall is used to store fabric, raw material like bales of wool from England and finished cloth, and it sensibly harbors scores of cats to clear the building of mice and rats. Except this is a bad day for these cats.

The felines may have succeeded in limiting the spread of vermin, but they themselves have had many litters over the winter, and now

Where Did Cats Come From?

it's Cats Wednesday. This is the day that the people of Ypres gather up as many cats as they can, walk them to the top of the Cloth Hall tower—70 meters (230 feet) up—and throw them over the edge to the ground below. Of course most of the cats are gruesomely crushed and shattered, but this day, one cat (according to eyewitnesses, the last to be tossed) survives the incredible fall and runs away to safety.* This lone survivor was symbolic: While Cats Wednesday had been an annual celebration, just one of countless atrocities against cats that characterized the Middle Ages in Europe, the 1817 version was the last time it was held. It's been replaced these days by Kattenstoet, or Cat Parade, a day of floats, costumes, and plush cats raining down on the crowds below. It's a fun day, though it falls short of atoning for centuries of felicide.

Cats Wednesday is one example of the many horrifying ways cats were persecuted in Europe centuries ago, but it was actually one of the least sadistic.

Cats became embroiled in a vicious circle involving the Roman Catholic Church, witches, and Satan (two of which are imaginary). That didn't stop Pope Gregory IX from announcing in 1233 that the devil was half cat (to be honest, I've known cats that qualify for that title). The papal proclamation claimed that anti-Catholic sects were conducting initiation rites in which a man would denounce his Catholic faith and would then be approached by a black cat—walking backward—and would kiss the cat where you normally wouldn't think of kissing it. Later the devil himself would appear as a man, "his lower part shaggy like a cat." Yes, that's a quote.

While the target of this bizarre document was apparently a variety

---

* Nothing more is known about this incredible survivor, except to say that physicists should celebrate it: It demonstrates once again that cats can use angular momentum to twist upright in midair, even if thrown upside down. And yes, it's been documented that cats can survive falls of 90 meters (roughly 300 feet), more than the height of the Cloth Hall tower.

of groups embracing unorthodox beliefs as they rebelled against the authority of the Catholic Church, somehow black cats and then witches were swept up by the church's enthusiasm for crushing perceived dissent.

It wasn't just Pope Gregory either. In 1484, Pope Innocent VIII declared that "the cat was the devil's favorite animal and idol of all witches." It was all about heresy, and as repulsive as it seems today, this campaign against cats not only lasted for centuries but flourished, limited only by the creativity of those conducting it. Anthrozoologist James Serpell put it most succinctly: "On feast days, as a symbolic means of driving out the Devil, cats, especially black ones, were captured and tortured, tossed onto bonfires, set alight and chased through the streets, impaled on spits and roasted alive, burned at the stake, plunged into boiling water, whipped to death, and hurled from the tops of tall buildings; and all, it seems, in an atmosphere of extreme festive merriment."[3]

Burning cats alive—especially black cats—was a big favorite. In celebration of the coronation of Queen Elizabeth I of England, an effigy of the pope (not Gregory IX) was marched through the streets. This wicker version of the pope was stuffed with cats, all of which were burned alive as the effigy was thrown on a bonfire.

Witches became a convenient stand-in for cats. It was popularly believed that witches appeared as gigantic cats, especially when engaged in evil. In 1582, an Englishwoman named Ursula Kemp was tried and executed for being a witch. Crucial testimony was provided by her son, who claimed she kept four animals as familiars, two of them cats. And in 1712, Jane Wenham, one of the last women in England to be condemned for witchcraft, faced an array of accusers who testified that her cats had tormented them and one of them had her face. Amazingly, she was pardoned.

Even if some accounts of the mass burning of cats are exaggerated, there's little doubt that cats, especially black cats, were targeted for

Where Did Cats Come From?

centuries in Europe. A typical story describes a greedy priest standing at the bedside of a rich man, hoping to capitalize on his death, while a deacon attends a poor widow. The deacon is surrounded by angels, the greedy priest by black cats.[4]

However, some argue that such persecution wasn't nearly as widespread as portrayed, and it's true that, even if it was, it apparently had no lasting impact, given the numbers of cats—even black cats—today.[5] But that doesn't mean wariness of black cats has disappeared. A 2024 study revealed that there remains some bias against black cats, particularly among people who are religious or who believe in witchcraft—especially if such people are primarily dog-lovers. But the best part of the study revealed that as Hallowe'en approached, undergraduate psychology students reported increasing anti-black cat feelings. It doesn't appear those biases were well grounded in fact.[6]

It's also true that while cats were being persecuted centuries ago, they were at the same time treasured pets for many. Even so, there was a vendetta against cats that dogs never experienced.

But why cats? Of course we have no direct access to the minds of citizens of the twelfth century onwards, but there are theories. One rests on the very nature of cats themselves, animals, even as pets, that are independent in spirit and not nearly as devoted as dogs appear to be. They couldn't be trusted, they snuck in and out of houses, and the females seemed extraordinarily determined to have sex (hence the association with witches and the devil). And black cats were particularly shadowy creatures.

Dogs, of course, were very different, and anyway, by the Middle Ages, they had established themselves as animals with defined roles: herders, hunters, or guards. Cats played the same role that they had always had: mousers. That was true ten thousand years ago and ten centuries ago. And today, for that matter.

Some scientists argue that this long-term consistency is exactly what should be most appreciated about cats. Unlike dogs, which have forsaken most of their wolf behaviors to establish close social relationships with us, cats have managed to do the latter—becoming valued pets—while at the same time maintaining their role as a predator: "Predator for Hire," as the title of one scientific paper put it.[7] The authors of that paper point out that some of dogs' social behaviors that we accept as part of the canine package, like a strong attachment to an owner, or separation anxiety, are much less common in cats. Cats have somehow added enough cozy-with-humans behavior to their repertoire while still terrorizing mice.

You might expect that as time passes and cats spend more time with us, they would become more attuned to us. John Bradshaw, in his book *Cat Sense*, wonders if the reverse might be true.[8] He argues that the majority of neutered cats are house cats, where the "wildest, meanest ferals are likely to escape the attention of cat rescuers and breed at will," meaning any genetically based tendencies to socialize with humans would face an uphill climb. Maybe that doesn't even matter: Most of the time, it looks like cats have already hit a sweet spot as the pampered hunter.

– CHAPTER 8 –

Costs/Benefits to Humans

> A small pet animal is often an excellent companion for the sick, for long chronic cases especially. A pet bird in a cage is sometimes the only pleasure of an invalid confined for years to the same room. If he can feed and clean the animal himself, he ought always to be encouraged to do so.
>
> —Florence Nightingale, *Notes on Nursing: What It Is, and What It Is Not*

I'm sure that the idea that somehow we are being parasitized by our pets, that they are taking advantage of us for their own good, seems absurd to most of you, pet owner or not. We love our pets! A dyed-in-the-wool evolutionary scientist might come back with "Yes, and yellow warblers love their cowbird offspring!" But as I've said, I don't feel completely comfortable applying Darwin's ideas to pets, given that the vast majority of them are sheltered from the natural world. However, it is still worth identifying exactly how pets benefit us.

The first step is to acknowledge that the benefits have changed dramatically over the centuries. The domestication of wolves to dogs tens of thousands of years ago was unique and world-changing, the first in a series of domestications that created today's agricultural world, populated by cattle, pigs, sheep, horses, chickens, turkeys, and . . . pets. You can't really measure the impact of all that. Some scientists have argued that without those animals, especially the ones raised for food and labor, the number of humans on the globe might be something like two billion, rather than pushing toward nine billion. Obviously, a great deal of that impact has been felt in the last two thousand years, and maybe even in the last few hundred, but it's also true that, at least on a scale of centuries, it could be that the impact of dogs began to be felt almost immediately after the first tentative associations between wolves and hunter-gatherers.

In her 2015 book *The Invaders: How Humans and Their Dogs Drove Neanderthals to Extinction*, Pat Shipman, an anthropologist at Penn State, makes a compelling case for the first dogs being a key element in the extinction of the Neanderthal people and the ultimate flourishing of us moderns. She takes an ecological view of the balance of predators and prey at that time, thirty to forty thousand years ago, and argues that the teaming of the two apex predators, wolves and modern people, allowed the pair to outcompete any and all other predators. Wolves earned their keep by alerting the group to the presence of danger, by contributing their superior vision, mobility, and sense of smell to the hunt, and the wolves that did that best would have been fed and allowed or even encouraged to breed. (Of course this storyline is more closely linked to the idea that wolves attached themselves to humans, rather than vice versa.)

But Shipman doesn't limit her analysis of the human-animal relationship to that admittedly speculative scenario. She also believes that, in general, domestication—and, by extension, the creation of pets—was driven by the adoption of animals as "living tools." Tools with a wide range of uses, including muscle power, rapid transport, wool or hair for fabric, manure for fertilizer, trash disposal (by consumption), high-fat and high-protein food, protection, and help in hunting. The combination of several of these could even have made it possible for humans to open up new habitats, as in the Arctic.[1]

Humans went on to be the dominant mammal on the planet. Dogs only made it to sixth place, behind rats (!), sheep, cows, and goats, but dogs are the most numerous pet. Cats are two spots behind dogs (just after pigs) in the mammalian rankings.

The thing to notice about these rankings is that only two of these mammals are commonly kept as pets, while the rest (except rats) are used for food. Sheep, cows, goats, and pigs remain as examples of Shipman's "living tools," and in fact, many dogs qualify on the same grounds. These include hunting dogs, guard dogs, guide dogs, herding dogs, search and rescue dogs, service dogs, and detector dogs. Depending on the culture, these dogs may or may not be treated as pets, at least in the sense that we're used to.

But there are cats and dogs (and others) whose treatment takes the idea of a "pet" to levels equal to those seen in royal courts of the Middle Ages. Dogs that are treated, well, like royalty. Once, when walking a friend's dog, my daughter encountered a woman in a long fur coat (!) walking her dog in the same park, and asked if she could give the woman's dog a treat. "Oh, no!" the woman said, practically cringing. "My dog only eats organic salmon." Stories like this help make sense of the finding that in the 86.9 million American homes that house a pet, 97 percent of the pet owners in those homes treat the pet as a member of the family.[2]

Sounds like another benefit to the pets, but surely we must be getting something out of it? We certainly do, but exactly what is the question. And it's a hotly debated one too.

The idea that having a pet could benefit the health of the pet owner has a long and well-attested history—there are many, many books built on stories along the lines of "When I lost my job and my best friend died, I was lost. I think the only reason I was able to recover was because I had [fill in the pet's name]. I owe my life to [him/her]." I have no doubt that these stories are true, but because they're personal, it's hard to figure out exactly what turned around the person's life. Was it mental, physical, some sort of combination of both, or even something that we might never be able to put a finger on? Obviously, these highly personal and sensitive experiences are not witnessed by an independent observer applying a battery of psychological and medical testing. Nor is it possible to confirm that the pet did indeed become more attentive in response to the owner's struggles, as many of these stories claim: It's well known that people attribute emotions to their pets that are more a case of wanting or expecting to see a reaction than actually seeing it.

Many of these studies claim that having a pet relieved depression, even to the point of persuading the owner to stop short of suicide. But . . . evidence? At the end of 2022, Harold Herzog of Western Carolina University updated a study he had done previously on the effects of pet ownership on depression. In his words, of fifty-one such studies: "35 of these found *no* differences in the depression scores of pet owners and non-owners, 7 reported that pet owners were more depressed, 6 reported those with pets were less depressed, and 3 found mixed results" (italics mine). Curiously, the more participants in the study, the less likely that pet ownership was to exhibit a positive effect. In Herzog's review, the eleven most populous studies showed no effect, while the studies revealing a negative effect involved more people than those showing a positive.[3]

It's easy to imagine why being with a pet would lessen depression—they're playful; as many attest, they offer "unconditional love"; they divert our minds—but it's more difficult to imagine the reverse. In this case, anyway, Herzog was unable to come up with an explanation.

This to-and-fro, evidence for and against, is not unusual in the study of benefits owners derive from their pets. It's a mixed bag of positive, neutral, and in some situations negative.

For instance, the modern era of scientific insights into the medical benefits of pet ownership kicked off in 1980, when Erika Friedmann and her colleagues published a study called "Animal Companions and One-Year Survival of Patients After Discharge from a Coronary Care Unit." Ninety-two patients who had experienced either a heart attack or angina—the chest pain associated with heart disease—were followed up one year after leaving hospital. Seventy-eight of the ninety-two had survived, but the crucial point was the disparity in survival between those who had pets and those who didn't. Only three of the fourteen who died had owned pets. To put it differently, 6 percent of the pet owners died versus 28 percent of the others. Recognizing that owning a dog often necessitates the beneficial activity of dog walking, the experimenters looked at those who owned pets other than dogs: There were ten, but none had died. The enhanced survival couldn't be pinned on dog owners walking their way to survival.[4]

The research team also found that "pet ownership could not be said to substitute for the beneficial effects of human contact"; pet owners with human companions still benefited from owning a pet. An interesting additional comment was that the owner-pet relationship does not necessarily require speech (an activity that raises blood pressure), which might also have been of value; however, I could also envision circumstances where saying some words to a pet might actually lower blood pressure.

This study is still cited today, forty-five years after it was published, a reminder that a scientific advance with impact can be both straightforward and brief.

Where has this research taken us in the years since 1980? There have been many studies, especially of the effect of pet ownership on blood pressure. The same Erika Friedmann published a study in 2013 measuring blood pressure in pet owners as they went about their normal day. Both key measures of blood pressure were lowered in dog owners in the presence of their dog, but only one in cat owners.[5]

The study scenarios are all very different. One small study that deserved attention because it was properly randomized measured blood pressure in a small group of thirty people waiting to adopt a dog. Half the group immediately adopted; half had to wait. The blood pressure readings in the two groups were equivalent before the experiment, but two to five months later the immediate adopters registered lower blood pressure. As soon as the delayed adopters acquired their dogs, their blood pressure fell as well.

There are many more blood pressure studies, but with a lot of variation among them, with some hard to replicate, others hard to generalize from.

In the article on dog adoptions, the authors enumerated the various impacts of pet ownership on heart health. They found, for blood pressure, four studies showing pets associated with lower blood pressure, two neutral, and one higher; for physical activity, the vast majority of dog owners walked more than nonowners; the only study involving cat owners showed that they walked 9 percent less than even those who didn't own a pet.* Most dog owners had higher or equal body mass index (BMI) than nonowners, and bird and cat owners tended to be overweight. Owning a pet did *not* lessen the risk of mortality, and reactions to stress were best summarized by this line: "Pets elicited the lowest reactivity to stress, whereas spouses caused highest."[6]

Mixed bag indeed! To summarize, I'd say dog ownership usually increases physical activity, which is healthy; blood pressure reduction is ambiguous, as is relief of depression.

A final comment about the mixed results that characterize so many studies exploring the benefits of pet owning. A recent analysis of twenty-four studies of pet owning, social isolation, and loneliness during the COVID-19 pandemic found uncertain evidence that pets reduce loneliness, and no major differences between dogs and cats. The difference here between isolation and loneliness is important: Isolation means literally that—you're alone, possibly by choice. Loneliness is different. You can be surrounded by people, but if they're not your people, you

---

* To muddy the waters further, one study showed that while dog owners walk more than cat owners, the benefits that accrue to their mental health come from walking *without* their dogs. I get it! I walk a dog who stops abruptly about every eight steps to smell something new.

can be lonely. In the twenty-four studies examined here, pet ownership appeared to have no effect before COVID-19 with respect to loneliness, but did afterward. Yet pets *did* reduce loneliness among adolescents before COVID. People who owned dogs fared somewhat better, which is attributed to the fact that walking the dog at least took people outside. Cat owners didn't do as well psychologically.[7]

It's complicated. Yet a third study published in 2022 ruefully concluded that "it is still a largely unanswered question whether pet ownership can reduce loneliness, or whether individuals who are more or less lonely may be likely to acquire a pet (or a combination of both)."[8]

But medical effects don't exhaust the benefits of pets. Not by a long shot. There are social issues too.

One that is more speculative than evidence-based but makes sense, at least at first glance, is that the attention paid to a pet might help establish or reinforce good instincts for parenting. This, of course, would be of some indirect benefit to humans intending to have children. It's very hard to find ways to provide scientific evidence for this, but there are indications that interest in pets is highest among girls aged eight to twelve, which *could* be seen as being aligned with this idea.

Closely related to this thought, and with some experimental grounding, is the claim that pet-keeping enhances empathy. One such study, which matched results on the "Animal Attitude Scale" with the "Interpersonal Reactivity Index," concluded that pet owners were more empathetic.[9] It's worth noting here that some benefits, if exaggerated, could be disadvantageous. A pet owner who is strongly empathetic may become anxious and frustrated by the demands of a pet, feel inadequate to care for it properly, and fear the grief that will almost certainly come with the pet's death.

And then there's oxytocin, the "love hormone," which plays a key role in birth and lactation, and also in reducing stress and anxiety. It's simple: When mothers and newborns gaze at each other, their oxytocin levels rise, which, in turn, facilitates bonding. The same thing happens when a dog and its owner gaze into each other's eyes, something that does not happen when a human gazes into a wolf's eyes. It's suspected this unique dog-human oxytocin connection arose during the domestication of the dog, possibly in two steps; first would have been the reduction of stress and fear among wolves initially exposed to humans, and the second building on that by strengthening social behavior. Genetic studies have shown there are indeed slight differences in the oxytocin receptor gene in dogs and wolves. The happy spin-off is that oxytocin exerts limiting effects on the brain's stress hormones, a benefit to us—and the dog.*

Does having a pet influence how other people perceive you? Evidently that's the case. Anyone who's walked a dog knows that having a dog beside you can spark conversation with passersby, although much of the conversation involves the passerby looking at, or even talking *to*, the dog. However, one French experiment took this idea a bit further. A young male named Antoine, casually dressed, was recruited to approach young women in a sunny outdoor park. Sometimes he had an average-sized mixed-breed dog with him; sometimes he was alone.

When young women eighteen to twenty-five years old happened along, he would approach them and say: "Hello. My name's Antoine. I just want to say that I think you're really pretty. I have to go to work this afternoon, but I was wondering if you would give me your phone number. I'll phone you later and we can have a drink together someplace."

With a line like that, Antoine would need all the luck he could get. But he did have good luck as long as he had the dog with him. After his invitation, he was to wait a few seconds. If the woman he'd approached turned him down, he just said, "Too bad. It's not my day. Have a nice afternoon" and walked away. (Another slick line.) If the

---

* The oxytocin connection is not limited to dogs! Those cats that are most secure with their owners also have higher levels of the hormone.

woman did give him her phone number, he'd immediately break the spell and explain the experiment. The results: With the dog, 34 out of 120 women handed over their phone number; without the dog, only 11 out of 120.[10]

So the dog is the magic, but apparently not just any dog.

In a similar experiment, experimenters recorded the reactions (smiles, hellos, or conversations) of pedestrians approaching a woman who was with one of three dogs, a Labrador retriever puppy, an adult Lab, or a Rottweiler. She also was occasionally equipped with either a teddy bear or a potted plant. The woman attracted little interest when she was carrying those, and while the two Labradors attracted lots of attention, the Rottweiler didn't.[11] Preconceptions!*

This isn't the only way that pets can attract attention to a human. The guy on the subway with his boa constrictor draped across his shoulders is another. A different pet with a different message.

It's really no surprise that pinning down the benefits of pets to humans is murky and complicated. Humans are complex and, as we're gradually beginning to realize, so are our pet animals. Whether the assumed benefits are as widespread as we might have thought doesn't seem to matter. The number of pets worldwide is growing nonstop regardless.

---

* This result prompts me to issue an apology. Nicolas Guéguen, the researcher responsible for the Antoine experiment, earlier designed a similar experiment showing that a young man holding a guitar, rather than a gym bag, prompted more positive responses from women. So I recruited my friend Steve Dodd, an excellent guitarist, to participate in a television version of the experiment for *Daily Planet*. Sadly, it didn't seem to work, but now I realize that just as it depends on the kind of dog, it wasn't Steve—it was the guitar!

– CHAPTER 9 –

Costs/Benefits to Pets

If you choose to evaluate success from an evolutionary point of view, where you rate success by numbers—that the more, the merrier—pets are doing fantastically well. Naturally, cats and dogs are the animals for which the most complete statistics exist, and they're pretty astounding. I call them statistics, but they are really only estimates, and they differ pretty substantially from one source to another. Nonetheless, there are features worth noting: the sheer number of dogs and cats, the steady upward trend in numbers, and the difference between pets that have homes and those that don't.

Accepting that sources vary, let's start with what I've found from a couple of websites that might be somewhat biased:

Dogster.com: "It's been estimated that as of 2022, there are more than 900 million dogs worldwide. . . . It's estimated that there are around 600 million cats worldwide. That's quite a difference!"

Catster.com: "It is estimated that there are around 600 million cats compared to only 400 million dogs."

At least they're in agreement on the 600 million cats. But how can there be such significant contradictions in the numbers? All it takes is a close look at how such numbers are gathered. For the number of pets that are actually kept and housed by humans, mailed surveys, door-to-door inquiries, and even random telephone surveys have been used. A separate set of data can be gleaned from the manufacture and sale of pet products—obviously their own very rough estimate.

These numbers, in turn, can be used as a basis for calculating the average number of pets per household. Multiply that by the number of households, and you have an idea of the total number of pets. But the uncertainties! People living in a "no pets" apartment are not going to report their three cats; if the survey is door to door, how to account for gated communities? What about people who just don't want to admit they have pets? It's even possible, I suppose, for people to claim they have pets when they don't. And are rural and urban areas equivalent? It's not difficult to see how significant inaccuracies would accumulate.

And that's just the number of dogs and cats *in* households. Those animals that are "village" or "free-ranging" or "feral" must somehow also be taken into account. Depending how carefully such data are gathered, and how fine-grained it is, we're getting into areas where uncertainties are almost unacceptably large. Still, when the differences from country to country and even from rural to urban are taken into account, the ratio of dogs to humans globally seems to be relatively constant: the more humans, the more dogs.[1]

And those unhoused dogs? I've read estimates there are, globally, 900 million dogs, of which 700 million are *not* living in homes. That's close to 80 percent of all dogs living on the periphery of human beings. Many of these dogs are still wholly or partially dependent on people, but some, termed "feral," live completely independently. A shocking number! Similarly, there are claims that 480 million cats, of the world total of 700 million, are *not* living in homes.

Another study suggests 700 million dogs globally, with 75 percent of those not having a permanent, human-supervised home. The absolute numbers are extremely variable, but they all point to a vast population of pets that don't live with people.[2]

The Unlikely Victors

But even this claim is disputed. Andrew Rowan, writing for Well-Being International, argues that there has been a steady recent trend away from street dogs to housed dogs, and the numbers might be something closer to 40 percent living primarily on their own, not 80 percent.[3] As people migrate more and more to cities, these numbers are bound to continue to shift.

I'm only quoting these numbers to support the idea that, regardless of living conditions—and in many of these cases, they are abysmal—the numbers of cats and dogs have gone through the roof since domestication. Wolves gave rise to dogs. Today there are several hundred million dogs compared with an estimated quarter million wolves. Add that only a few tens of thousands of wildcats survive, ancestral to the hundreds of millions of house cats, and, at least from the evolutionary point of view, the case is sealed for domestication being beneficial.

Reproduction is the machine of Darwinian evolution, and these most populous pets seem to have won that battle. Just to review: Dogs are the most common carnivore in the world; house cats are vastly more numerous than all forty or so species of wildcat, from cheetahs to lynx; and even chickens (the vast majority of which are definitely not pets) weigh more than the total of every other bird living today and are likely the most populous bird ever.

But these are examples of human intervention. Not spectacularly successful evolution. Yes, it's selection, just not *natural* selection. It doesn't really make sense to look at the burgeoning numbers of our pets and label that a masterpiece of evolution. You could argue that they coped with a dramatically changing, human-dominated environment via domestication, but they didn't do it on their own.

Even if this evolutionary approach isn't directly relevant, it's still worth pursuing the question of which side, human or pet, has benefited the most. While booming populations of pets are undeniable, quality of life is a different story.

For one thing, not all cultures treat their pets in the same pampered way commonly seen in pet food ads or TikTok and YouTube videos. It's important here to draw the line between animals domesticated for their economic benefits and animals that aren't, although a handful of species can play both roles—there are always exceptions! On the pet side, there are hamsters, rats, guinea pigs, rabbits, tropical fish, budgies, parrots, monkeys, snakes, reptiles, and house cats. Other than cats and their superior ability to eliminate mice, the economic contribution represented by the others is negligible. Then there are animals that are almost entirely (not absolutely) kept for their usefulness or monetary value, including oxen, cattle, llamas, vicuñas, pigs, and sheep. On the borderline, there is a select, but short, list of animals such as chickens, horses, and dogs.

The borderline animals not only have dual roles, but can be treated very differently depending on the role they're playing. Thoroughbred horse racing has been plagued by deaths of horses and charges of drugging, though at the same time, it's not difficult to find articles extolling the exemplary way Thoroughbreds are cared for. The format of chuckwagon racing in rodeos has had to change to reduce the number of horse fatalities. But there are also horses that are considered members of families.

Broiler chickens sadly have been singled out for having, in the worst-case scenario, a space no bigger than an 8½-by-11-inch sheet of paper to live their entire lives in, but again there are backyard chickens that live—as far as we can tell—contented lives. Harold Herzog, in his book *Some We Love, Some We Hate, Some We Eat: Why It's So Hard to Think Straight About Animals*, makes the case that a rooster raised for cockfighting, which, after being pampered for many months, likely dies a gruesome death, ripped apart by the metallic spurs of its opponent in a cockfight, still lives a better, longer life than the broiler chicken that never sees the light of day.

A quick look at human-animal relationships around the world presents a picture that might surprise you, in terms of both the

The Unlikely Victors

menagerie of animals treated as pets and the way they're treated. Take North America, where it's fair to say the focus on pets as companions is more intense than anywhere else in the world. In the United States, in 2022, Americans spent $136.8 billion on their pets; given that number is rising about 11 percent per year, by now it's likely pushing $170 billion every year. This is a first-place position in the global rankings, but even Canada, with a population one-ninth that of the US, spends close to $8 billion a year on pets. But this is a warped view of pet-human relationships that the rest of the world doesn't share, especially those parts of the world where the pet population is derived from the local fauna. The research on cross-cultural pet ownership makes crystal clear two important points: the incredible variety of animals labeled as "pets" (something I just touched on in chapter 3) but also the wall-to-wall spectrum of treatment—good and bad—of these pets.

You'd practically need a copy of an Encyclopedia of Mammals, Birds, and Reptiles to identify them all, but a short list of world pets includes macaws, sloths, capybaras, coatis, caimans, bears, ostriches, tortoises, opossum-rats, lowland pacas, agoutis, bats, tapirs, peccaries, ocelots, and margays.

I'm now convinced that the number of wild animals that haven't been adopted as a pet, somewhere, by someone, is very, very small. (Not surprisingly, in every place where you find unusual pets, there are often house cats and usually dogs around as well.) Francis Galton, who is responsible for some of the most intriguing early observations on global pet-keeping traditions, claimed "that Mr. Bates, the distinguished traveller and naturalist of the Amazons has favoured me with a list of twenty-two species of quadrupeds that he has found tame in the encampments of the tribes of that valley."[4]

Ah, but being a pet in this vast array of circumstances isn't always the treat you might imagine. The treatment of pets is as much a local custom as the choice of species. So, for instance, when young animals are adopted, it's not uncommon for them to be killed—sometimes eaten—when adult.

You might assume that what North Americans consider mistreatment is peculiar to the specific animal, but when Peter Gray and Sharon Young of the University of Nevada at Las Vegas analyzed cross-cultural data for dogs, they found it's not so much the animal; it's the humans.[5] Of the sixty cultures surveyed, dogs were kept in fifty-three, but only treated as pets, in the sense of companions or family members, in twenty-two of those. In the rest, dogs were workers, helping with hunting, herding, or defending their owner against other humans or predators. Of the twenty-two cultures who treated their dogs as pets (at least in this definition), a mere seven allowed the dogs inside. That's fewer than those where dogs were beaten—thirteen—or killed—eleven. It's not the animal.

But is it always just a case of wanton cruelty? Not necessarily. Often, as a young animal becomes a more aggressive adult, or is now unable to provide the services it used to, its human caregivers eliminate it, but as a custom, not a cruel act. For instance, the Sami of the Scandinavian countries, also known as Laplanders, have been herding caribou for thousands of years. According to one anthropologist, Hugo Bernatzik, in his 1938 book *Overland with the Nomad Lapps*, when a Lapp dog was no longer able to keep up with the demands of herding, it was to be hung from a tree, but the owner would never do it—he would ask a friend instead.

In 1978, the anthropologist Merrill Singer wrote about the BaMbuti Pygmies of what was then Zaire (now the Democratic Republic of the Congo). They were dependent on their dogs to help in the hunt but treated them with aggression and disdain. Singer quoted the anthropologist Colin Turnbull as saying the Pygmies' dogs were "kicked around mercilessly from the day they were born to the day they die." And yet their dogs were crucial members of the community, essential to the tracking and killing of game. And, to add to the mystery of the cruelty,

The Unlikely Victors

dogs injured in the hunt were treated with care. Singer could only conclude that in this tight-knit community, where cooperative hunting was essential to survival, aggression directed at humans would be too disruptive, so it was directed at the dogs instead. Would it be surprising that mechanisms like this can underlie cruelty to pets everywhere in the world?[6]

It's not just the welfare of pets that's dependent on the cultural practices of their humans, but even the choices of pets themselves. And you don't have to do a worldwide sweep to see that. Harold Herzog enumerated changes in pet popularity over centuries in Europe alone, from monkeys in the thirteenth century through tortoises and hedgehogs a couple of centuries later to the so-called pocket pets like mice and toads in the eighteenth century.

Similar crazes erupted in the United States for caged birds and aquarium fish at the beginning of the twentieth century, and in Japan for beetles in the 1960s.

It seems that for a pet to truly benefit from its association with humans, it needs to be the right species at the right time in the right place. And rarely does the pet get to choose.

– PART III –

Sorry, We're Out of Dogs and Cats, But . . .

– CHAPTER 10 –

Horses

The world of pets is dominated by dogs and cats. But their reign extends to all of nature. In the ranks of all the world's carnivores, the dog, 900 million strong, and the cat, 600 million, outnumber all others. The third-place animal, the harp seal, has a mere 4.5 million. *Canis lupus familiaris* and *Felis catus* are ubiquitous, our relationships with them personal and emotional. There are developing scientific accounts of their origins, their social behavior, and their intelligence. No other species of pet is known inside out like dogs and cats.

Still, every pet has some attribute that attracts human attention, maybe even affection. These next few pages are organized differently. Four different pets, three of them already classified as pets (though nowhere near the popularity of cats or dogs) and a fourth that is at least potentially a pet, but may never be. Nevertheless, it and the others each have some unique feature that makes them intriguing. As usual, the intrigue, whatever it is, sheds more light on the pet owner than the pet.

One animal that has a long history of relationship with humans, yet seems an awkward fit with the label "pet," is the horse. But Carolyn Willekes, a classical scholar/horse-lover of Mount Royal University in Calgary, Alberta, argues that there is very early evidence that horse and human can have that empathetic, emotional relationship that characterizes pet owning: the famous anecdote of Alexander the Great and his warhorse, Bucephalus.

As is often the case with tales from ancient Greece, the best sources for a story are those written centuries after the event happened. That's the case here. The Greek historian Plutarch wrote about Alexander four hundred years after his death. Born in 356 BCE, Alexander succeeded his father, Philip, to the throne of Macedonia when he was twenty. He then embarked on a series of military adventures, conquering territories as far away as northern India that together formed one of the largest empires in history. He was undefeated in battle and, Plutarch contends, rode Bucephalus in all of them.

When Alexander was twelve or thirteen, his father was negotiating to buy this hugely expensive horse, but it was too skittish—it could not be "broken" to accept riders.* Alexander stepped forward and claimed that he could train Bucephalus when all others had failed. He then demonstrated an understanding of horses that seemed to have escaped those others: He noticed that the horse startled when he saw his own shadow, so he kept Bucephalus facing the sun, and the horse calmed. Some versions of this story claim he also moved a distracting scarf that was lying on the ground. But the important point is that Alexander shunned brute-force techniques and worked with Bucephalus to establish a relationship with him. The Oracle at Delphi had told Alexander's father that whoever rode Bucephalus would rule the world. As usual, the Oracle was right.

Professor Willekes's response to the story of Bucephalus—"Although dogs are traditionally considered to be our best friends, it is upon the back of a horse that human history was written"—makes the case for the importance of horses to humanity. But horses as pets? Most of us just don't think of them that way. Their role in history is undeniable,

---

* A good rough estimate of Bucephalus's value (apparently 13 talents, back in the day) might be the equivalent of nearly seventeen years' pay for a Greek worker.

but they live outside with the rest of the livestock, not curled up on the couch. Yes, they are pets for many people, but there's an uncertainty around them. Are they working animals? Are they warriors? Athletes? Yes, all of those, but why don't we think of them as pets when many horse owners have a deeply emotional relationship with their animal? Is it because you can't really keep them in your city home,* they're intimidating, they don't seem as responsive as, say, dogs or cats, and they are very, very big? They've been called the "in-between" pet.[1]

I don't have much experience with horses, but I have quickly come to admire their uniqueness, their special qualities, and their impressive history.

The domestication of the horse is a little more mysterious than that of the cat or dog, and the time frame is somewhat unclear. Just as I was writing this, the estimated date for humans beginning to ride horses was revised to be a thousand years more recent than previously thought.

For context, dogs could have been domesticated 30,000 years ago, cats about 9,500 years ago, and sheep, goats, cattle, and pigs all between 10,000 and 6,000 years ago. Horses, though? Up until 2024, the accepted time for horse domestication was about 5,000 years ago, but it has now been shifted to a more recent 4,200 years ago—much more recently than all those other animals.

Braiding the different strands of evidence together is what makes dating the domestication of the horse challenging. There has been a long-standing belief that the hordes of Yamnaya people (speaking a language ancestral to the Romance languages) who swept across Europe and Asia five thousand years ago could have moved with such apparent speed only if they had horses—that is, domesticated, rideable, milkable, edible animals—under their control.

---

* Although Shirley Watts, the wife of the Rolling Stones' drummer, Charlie Watts, brought an Arabian horse (she bred them) into the drawing room of their English home for a *Vanity Fair* photo shoot.

Sorry, We're Out of Dogs and Cats, But . . .

That might make sense, but where's the evidence? In 2023, an analysis of Yamnaya skeletons revealed a small number whose hip bones showed the kind of wear that constant horseback riding would produce. But the skeletons were not in pristine condition and therefore not the strongest case possible for having domesticated—and ridden—the horse five thousand years ago. People who follow, or participate in, this research urged caution for that reason.

It's really that number, five thousand: a well-attested date for the Yamnaya's move into Europe. If they were riding horses then, those horses would have had to be domesticated five-thousand-plus years ago. Yet there has always been different evidence, solid evidence, that it happened more recently than that.

One is a burial called "the charioteer," a male human three thousand years old, found in Siberia. A length of metal with hooks at each end was found lying across the body. Similar devices have been found in Chinese burials. The hooks allow charioteers to throw the reins over them to free their hands. This burial is the earliest—and only—presumed charioteer ever found in Siberia. But if horses had been domesticated two thousand years before this, earlier charioteers should have been found—a mere three thousand years ago doesn't fit with the Yamnaya.

Then there are horse images, also around three thousand years old, on document seals or terra-cotta found in what was southern Mesopotamia. Again, not as old as the Yamnaya.

Then, in 2024, another difficulty for the five-thousand-year claim: DNA analysis of the remains of 475 horses that lived throughout this period found no genetic evidence that horses were domesticated five thousand years ago or more, but did identify something of significance around 4,200 years ago: the sudden appearance of a new genetic strain of horses, one closely related to that of horses today, together with an apparent shortening of the reproductive cycle from about seven years to four years, a sure sign that humans were meddling with what had been the natural generational pace. The study also revealed that this genetically novel horse spread rapidly across Europe and Asia, suggesting that finally, by this time, horses were being ridden, rather than just being kept for milk and meat.

Of course this new data argue that the Yamnaya, who, in their epic migration, covered vast distances in a century or two, were trudging, not riding, an idea that still doesn't sit well with some.

It could be that horses were indeed managed earlier than that. However, once humans started riding them, a whole new life for them and their humans began. The new mobility set human history on a completely different course.

For instance, it wasn't long before horses were immersed in warfare. It can be argued that, despite the gore and violence horses were exposed to in battle, it was that role as warrior horse that called attention to an emotional, loving relationship between humans and horses that argues for their inclusion as pets.

It isn't just the legend of Bucephalus, although the fine detail of that saga stands out, even if it's possible that it is slightly embellished around the edges. It is said that not only did Alexander refuse to ride another horse, but Bucephalus refused to be ridden by anyone other than Alexander. They traveled—and fought—together over more than five thousand kilometers (three thousand miles). He named a city after this horse! And remarkably, 2,300 years later, in the First World War, a little more than a hundred years ago, horses were still a crucial piece of any army. They're no longer instruments of war, but they were for this entire stretch of time, from ancient Macedonia to twentieth-century England.

Belonging to a cavalry is, of course, not the same as being a pet; relationships were being formed in the unusually horrid conditions of battle, but their intensity reveals deep emotional ties.

We know something of the human side of such a relationship from First World War soldiers. Hundreds of thousands of horses fought in the Great War. Before war was declared, the British Army possessed 19,000 horses, but 53,000 arrived in France in August 1914. By August the following year, the British Army had more than half a million horses, many plucked from British farms, but also thousands imported from other countries. It's hard to picture from today's vantage point just how essential they were: The biggest of them towed artillery pieces; the sleeker ones carried soldiers.

Horses were essential cogs in the machinery of war. It was clear that when the British Army decreed that soldiers were responsible for their horses' well-being, this was as much—or even more—about the horse's abilities to carry out its duties as it was out of concern for the horse per se.

True enough, but there was something more: "Horses may have been weapons of war, but they were not hairy motorcycles," Jane Flynn has written.[2] Inside the army and out, there were endless stories of close relationships between a soldier and his horse. As Flynn, the author of *Soldiers and Their Horses: Sense, Sentimentality and the Soldier-Horse Relationship in the Great War*, points out, stories like this played well with the British audience, who liked to think of themselves and their countrymen as brave *and* compassionate.

The magazine *The War Illustrated* led the way in these stories, and they were plentiful enough to believe that there was truth in those words: "A great sympathy exists between cavalrymen and their chargers, and there have been many instances of horsemen with tears in their eyes, giving their wounded animals a fond caress, and then putting them out of their agony."[3]

Indeed, in July 1916, *The War Illustrated* depicted a heartrending scene on the cover, an incident where a young soldier's horse had been mortally wounded. The soldier, rather than escaping from pressing danger, held his horse's head in his arms and wept. And a horse's death wasn't even the worst; the worst was when the soldier himself had to shoot his horse after it had been fatally injured.[4]

Horses

Horses in war are even more extraordinary when you factor in that they are prey animals, not predators. Their lot in life is to avoid being killed and eaten; always has been. That is their evolutionary background. It's impossible to imagine what goes through a horse's mind in the middle of the ghastly, grinding theater of war, but their soldiers likely had a sense of what the animals were going through. Even in the relative calm of day-to-day life on a farm, a horse begins with the default of fear, and proceeds from there.

Some are of the opinion that fear is even apparent in the billion-dollar industry of Thoroughbred horse racing:

"Photos of racehorses snapped in the moments they extend full stride across the finish line show expressions of animals in fright," reported *The Guardian*. "The tendons in their faces are corded tightly, their ears are positioned in ways that indicate frustration and confusion, their eyes are wide in panic, and their mouths, dry from anxiety, are handled violently by the bit. This is the opposite of what horses want."[5]

A Polish team of scientists has recently detailed the fear responses of horses in situations that suggest these fears simmer just under the surface, and appear to derive from the animal's evolutionary history. In one case, horses were walked around an arena and encountered first a cow in a neighboring pen, and then an electrically powered mobile cardboard box one meter (three feet) high. In both cases, the horses' heart rates spiked, and they stayed well away from both box and cow, but to the surprise of the experimenters, their horses reacted more strongly to the cows and, curiously, reacted to one of the two experimental cows much more than the other (it wasn't clear why).

As the scientists put it in their report: "Horses, as prey animals whose initial reaction to potential threats is often flight, are sensitive to (potentially) all unknown frightening stimuli, and this propensity . . . *still exists despite years of domestication and selection against fearful behaviour*" (italics mine).[6]

The same researchers underlined that last point with a second study. They played the sounds of predators to different breeds of horses, attempting to match a horse with a predator sound it might have evolved to fear.

Their experimental subjects were Polish horses—Konik Polski—and Arabians. The horses heard the sounds of wolves, leopards, and golden jackals (as a control group), and there were different reactions. The Polish horses alerted most to the wolf vocalizations and came together in a tight group in response. The Arabians, on the other hand, didn't pay as much attention to the wolves, but, perhaps strangely, lined up facing the source of the sound of the leopard and approached it. It is tempting to argue that the different breeds arranged themselves in these different formations to suit the predators (wolves in packs, leopards operating solo), but the scientists admitted that was "highly speculative."

These are just two experiments of many, but as the Polish research group asserted, "The tendency to react with fear is deeply rooted in the equine personality and very difficult, if even possible, to eradicate. The fearfulness . . . has been explained as basic, genetically imprinted response enabling threat avoidance."[7] The fear of anything new or unknown helps makes some sense of the legend of the mighty Bucephalus startled by his shadow.

The horse, then, is an animal that approaches any novel interaction with fear as a motivating factor, and yet, as the war reports show, can form extremely close bonds with humans. To make the case that a horse fulfills all the emotional and intellectual criteria for a pet, it's worth exploring the human-horse relationship more closely.

Temple Grandin of Colorado State University, the famous autistic advocate for animal welfare, followed closely on Alexander the Great by identifying small, unexpectedly distracting objects, like a metal ladder or a raincoat flapping in the breeze, as fear-provoking issues for cattle—remove them, and the animals calm. She also had a lot of experience riding horses when she was young: "Riding a horse isn't what it looks like: it isn't a person sitting in a saddle telling the horse what to do by yanking on the reins. Real riding is a lot like ballroom dancing or maybe figure skating in pairs. It's a relationship."[8]

Contrast that statement with the traditional meaning of "breaking" a horse, which essentially meant breaking its spirit or will so that it would accept a saddle and rider. Tying its legs so it can't flee and using a whip were typical. Yes, it worked in the sense that the horse would come to accept that it was subordinate to the trainer, but the more we know about horses, their emotions and their intelligence, the less likeable such techniques are, and they have been replaced by what's sometimes called a "natural" approach. Think *The Horse Whisperer* (although the popular movie included some things that even "natural" horse trainers would never do, like driving the horse to exhaustion).

Aside from training, what is the nature of the horse-human relationship, an important piece of the picture for some pets? Humans equivocate: One Canadian survey of what were called "horse enthusiasts" found that a strong majority felt that horses could experience "frustration, depression, sadness, jealousy, anger, happiness and love" and considered them "companion animals," yet more than half still labeled them as "livestock."[9]

But Temple Grandin's "it's a relationship" has to cut both ways. Lately, that idea has been reinforced by scientific evidence of what's going on in a horse's brain, and the research supports the idea that a horse could be as much of a pet as a dog or cat if it were allowed to be. We need to examine some of the results patiently before drawing conclusions. For instance, in one experiment, horses were trained to touch a card with their nose to get a treat. When the task was made more complicated—touch the card only when an accompanying light was off—they kept touching the card regardless, light off or on. Sounds like abject failure, but then the experimenters dialed it up even further: If a horse touched the card when the light was on, it was given a ten-second time-out. Suddenly most of the horses waited until the light was off. The immediacy of their new response suggested to the researchers that the horses had understood the game all along, but didn't see any point holding back as long as there was no penalty. They're smart.[10]

Also, they're attuned to us in a way that's similar to dogs. This shouldn't be too surprising really. It's clear that horses recognize each other and maintain bonds on that basis. We're not horses, but if they're shown pictures of people with angry or happy faces, then confronted hours later with one of the two actual humans wearing a neutral expression, they engage in more stress-related activities like scratching and sniffing the ground when the human who had exhibited the angry face enters the room. The eye the horses favored when viewing either the angry or happy face suggested they were using the brain's left hemisphere to process happy faces and the right hemisphere when viewing angry faces. This correlates nicely to dogs wagging their tails left or right (see chapter 32). They're tuned to our facial expressions, and may even be tuned to our brains. Electroencephalograph recording of humans grooming and riding horses showed that the brain wave patterns of both horse and human tended to synchronize as they spent time together. Underlining this scientific evidence are the testimonies from people who interact with horses and describe innumerable ways a horse displays companionship: approaching, nuzzling, following, touching—everything a dog or cat would do or more.

People who spend time with horses come away with the impression that the animals are thinking and perceiving in ways we're usually not aware of. Some say horses study us and find us curious and amusing.

Speculative fiction writer (and rider) Heather Clitheroe told me a story of being on a competitive trail ride, in which horse and rider must follow subtle signposts to complete a course on time. The rider has a map, but the horse has no prior knowledge of the route whatsoever. It's not a race but more like the equine equivalent of a car rally:

Horse and rider must complete the course accurately, neither too early nor too late.

On one such competition, Heather and her horse came to a fork in the trail, and she urged her horse to go left. Her horse stopped, obviously unhappy with her choice. She persisted, but then, after a short distance, she realized she had been wrong and reined in her horse. It then gave her a definite side-eye look of impatience, an "I knew it all along." And it had.

Science fiction writer/horse-lover Judith Tarr, in thinking about communicating with animals, argues that the idea of "talking to the animals" is a mistake. It's not words—it's touch. Most of the time, horses don't kick or bite; they nudge. "Horses are even more expressive with their bodies. Their whole world is movement. They live in herds, where every individual is aware of every other. Humans can't come close to that physical or spatial awareness. If a horse is 'inferior' to us because they can't form human speech, a human is just as much so on the deeply physical level."[11] Tarr has also observed that humans in the presence of horses do a poor job of paying attention to what the horses are telling them: "The horses will be swirling around each other in ways that to them are eminently clear and, in body-language terms, extremely loud, with neon flashers, but the visitors are oblivious. They're all up in their heads with the words and the ideas."[12]

Horses possess every quality a good pet needs, plus a few that are bonuses, and they always impress with their strength, their speed, and their intelligence. But their size will likely always confine them to being pasture pets. Even so, horses get my vote as the best animal for pethood that hasn't quite made it yet.

– CHAPTER 11 –

Parrots

Horses are wonderful animals that do require considerable space and care, but if that can be provided, they definitely fit the bill of a pet. There is no ambiguity about parrots being pets—many of them are. But parrots are trickier than dogs and cats, and are a perfect example of how different animals attract different kinds of pet owners.

The three most famous parrots are Captain Flint, Alex, and the anonymous Norwegian blue from Monty Python.

These three parrots had very different owners. Captain Flint was Long John Silver's parrot in Robert Louis Stevenson's novel *Treasure Island*, best known for saying "pieces of eight." Alex the grey parrot belonged to researcher Dr. Irene Pepperberg, a scientist at Boston University, and the Norwegian blue was being returned to the pet store, unowned and uncannily silent. There are good reasons for their fame.

Treasure Island's Captain Flint, though fictional, was a bird that could easily have been spied on a pirate's shoulder, especially in the Caribbean, from the early 1600s to the mid-1700s. Piracy was having its moment in those days, and while actual treasure was the goal, loud, brightly colored parrots enhanced the style of any pirate. Not only that, such birds could be sold in Europe for solid prices. The value of the guilder fluctuated pretty dramatically at that time, but it's been estimated that a single parrot could earn a pirate 60 guilders. Depending exactly when that happened, it could be more than a year's earnings.

They were popular and expensive because they were eye-catching—if they weren't, people wouldn't walk around with them perched on their shoulder.

And they're also pretty smart. Alex the grey parrot was Irene Pepperberg's research subject for much of his thirty-one years, and seemed to have extraordinary language skills, maybe even conversational. It's worth remembering, though, that these feats were displayed under tightly controlled experimental conditions, not just sitting around chatting casually at home.

And the Monty Python "Dead Parrot" skit? I've watched it; it's hard to learn anything about either the bird or its owner.

These three elements—the look of the parrot, its smarts, and the commerce around it—are all issues influencing the bird's popularity and suitability.

In this chapter, I'm concentrating on large parrots, like the African grey (famous among linguists and cognitive scientists), and the sulphur-crested cockatoo, not small ones that can share a cage, like budgies or parakeets. No prejudice, but the large parrots engage in more behaviors that either challenge or attract pet owners, language being the most important.

One thing I am *not* going to do is to advise against owning a parrot for reasons like these published on the In Defense of Animals website: "Parrots are one of the most frustrating, destructive, messy, and noisy companions a person can have. This increases the odds that the birds will be abused and neglected, and finally rehomed, possibly with an even worse guardian."[1] Nor am I going to promote owning a parrot for these reasons: "Birds make great pets! They are playful, cuddly and sweet. Unlike dogs, they don't have to be taken for walks or played with. Most are content just hanging out on or with their favorite person."[2]

So many pluses and minuses: Yes, they can talk! But they can also outlive you!

Perhaps the most serious downside is the risk that when buying a parrot, you might be acquiring ownership of a bird that was captured in the jungle. I checked out the International Union for Conservation of Nature's assessment of the health of parrot populations worldwide. Of the 374 parrot species listed on the IUCN list, 232 are in decline, 107 are stable, and only 35 are gaining numbers. Statistics like that demand due diligence when buying a pet parrot.

Parrots are unique in that they are the only pet that you can converse with using the spoken word. This raises so many questions. If they are thinking about what they're saying, would we, one day, be able to get some insight into that?

Although parrots have apparently been pets for thousands of years, early records of their behavior are pretty sketchy, but we have to assume they've been talking for most of that time. The earliest reference to a talking parrot dates to 397 BCE. Setting aside language for the moment, many species of birds other than parrots are able to mimic sounds in their environment, whether that's baby birds imitating their parents, or adults imitating the calls of other birds or even random sounds like the swinging of a rusty gate or a car alarm. But this is imitation, not communication, unless these sounds have a role in the ongoing interplay among birds. And these mimicked sounds very rarely constitute human speech.

Parrots, on the other hand, are able to articulate words clearly enough that the most skilled can hardly be distinguished from a human. But do the birds consider the topic, make a decision about what words fit the context, and then say them? That would put parrots in a league of their own, not just among birds but among all animals.

Why do parrots have this outstanding ability to imitate or actually invent words?

It's all about the brain and the behaviors that brain makes possible. One study of parrots called orange-fronted conures has shed some possible light on the behavioral side. Wild-living orange-fronted conures have individual contact calls, and in the wild, it's been shown that each bird in the flock is able to imitate another bird's contact call virtually as soon as it hears it. Scientists tested the response of orange-fronted conures to others' contact calls and concluded that imitating a bird's contact call catches that bird's attention. In a situation in the wild where these flocks are in a never-ending cycle of breaking apart and then coming back together, using the contact call of a specific individual might be the easiest way to find that bird.[3]

Betty Jean Craige, owner of a female African grey named Cosmo, recorded what seems to have been a domestic version of that. Cosmo's utterances were taped whether Betty Jean was in close contact or not: in the room conversing with the bird, in the room but ignoring her and talking to another human, not in the room but within hearing distance, or out of the house entirely. It seemed Cosmo made many more word-like sounds when her owner could hear. Cosmo wasn't just making random noises, but turning what she said to conversation and, when appropriate, voicing what could be taken as "contact calls," like "Where are you?"[4]

On the brain side of the coin, the genomics of parrots have begun to be sorted out, and there are some startling discoveries, even if you're already aware that parrots are pretty fantastic animals. The parrot line has acquired or developed many new genes since they split from the line leading to songbirds. Some of these genes contribute to the parrot's incredible longevity (up to sixty years, sometimes more), while others influence the neural circuitry of the brain, especially in areas devoted to articulating verbal sounds. These same circuits have evolved in a parallel direction in humans. It was already known there were physical similarities in humans' and parrots' brains in those areas, and now it's clear that parrots are genetically tuned for this. The geneticists call it convergent evolution, where two vastly different species land on the same evolutionary answer to a challenge. In this case, it's shaping the ability to use specialized sounds—words—to communicate. Even though the brain of one of these is about the size of a walnut (the parrot).

Sorry, We're Out of Dogs and Cats, But . . .

It was really Dr. Pepperberg and the African grey named Alex who put the idea of language on the table.

Unfortunately, Alex died in 2007 at the age of thirty-one—young for a parrot—and while it's tempting to argue every linguistic ability he showed could generalize to other parrots, it's also possible that he was one of a kind, although that seems doubtful to me. Alex's detractors (I guess they were really Dr. Pepperberg's detractors) argued that maybe the parrot was getting unconscious cues to the answers from Dr. Pepperberg, who was usually sitting right in front of Alex.

I have a duty here to refer to Clever Hans, a German horse in the early 1900s that was claimed to be able to add and subtract, read and spell, and identify colors.* Hans indicated his choices mostly by tapping a hoof, and he was pretty skilled at those questions too, until it was shown that in the absence of his owner, or someone who knew the answer, Hans was no more skilled than any other horse. He had been picking up cues from his owner, Wilhelm von Osten. Turned out von Osten hadn't been trying to cheat: The cueing was unconscious, a slight tilt of the head, pursing of the lips—who knows?

No such cueing has ever been found in this case, however. The persisting caution surrounding Irene Pepperberg's experiments with Alex is generally based on a reluctance to claim too much on behalf of the bird, arguments along the lines of "He can name objects, but does he really know what they mean? Can he create a sentence using them?" Or "He names things, but only when they're there in front of him. Thinking involves talking about things which aren't there."

Here are some examples of what he *could* do: Alex learned to identify fifty different objects, seven colors, five shapes, numbers up to eight, and concepts like bigger and smaller, same and different. With a set of differently shaped, different colored objects made of different

---

* "Duty" because every time the issue of unconscious cueing comes up, Clever Hans is cited as the example. Every single time.

materials, he could answer questions like which are the same color, the same material, are these two the same or different?

At the same time, Alex liked to play around, sometimes deliberately offering the wrong answer. Of course, if he utters the wrong answer, it might be because he doesn't know it, but it could also be trickery. We encountered that while recording Alex for a story on *Daily Planet*: He'd throw in a wrong answer every once in a while, and Irene Pepperberg, who obviously knew Alex very well, felt some of them were deliberate.

My favorite Irene Pepperberg example of something like that came when she was doing a BBC radio interview. Holding a square orange piece of wood in front of Alex, Dr. Pepperberg said, "Tell me what color." Alex: "You tell me what shape." Dr. P.: "It has four corners. What color?" Alex: "Tell me what matter?" Dr. P.: "It's wood, can you tell me what color?" Alex: "How many?" Dr. P.: "There's only one!"

Pepperberg decided to take a break to interrupt this craziness, and as she walked away, she heard, "I'm sorry. Come here. Orange."

And then there was a demonstration involving colored letters, where Alex simply persisted in saying "Want a nut" every single time, until he lost patience and said, "Want a nut! Nnnn . . . uh . . . teh!" Dr. Pepperberg pointed out Alex had been trained to say *N* and *T* but not *U*. He came up with that on his own.[5]

Alex is definitely not the only parrot to have demonstrated word and reasoning skills, but he was the one that started to break down the barrier to even imagining that birds could think. Much of the research since his death owes a debt to him. And there has been some pretty cool work.

African grey parrots like Alex, described as the "titans of alien intelligence," aren't just smart; they're engagingly social. They are eager to share treats with other parrots—for no apparent reward—and if shortchanged display no anger or envy. Contrast that with, say, capuchin monkeys, which, if they get a piece of cucumber while the next-door capuchin gets a grape for doing the same task, throw an absolute tantrum.[6]

Likely the parrot experiments that have attracted the most attention are those comparing the birds' linguistic development with that of human infants. Two such studies, both published in 2024, reveal intriguing differences between the species.

One study focused on what's called "private speech" (or "talking to yourself"). Private speech is used by young children to practice and perfect spoken language. The use of private speech declines as children become more accomplished. It's also used by parrots. A girl named Emily had her private speech recorded from the time she was two years old until her third birthday, 122 private speech events in that time. (It should be noted that she was considered verbally advanced for her age.)

For comparison, the African grey parrot named Cosmo I mentioned earlier was recorded for nearly a year when she was six years old. Side-by-side comparison revealed many differences—no surprise there. But the differences are intriguing: Emily had a much bigger vocabulary, although the researchers pointed out that Cosmo's owner used very simple phrases when speaking to her parrot, so Cosmo's repertoire could be expected to be less. Emily also used connecting words—prepositions and conjunctions—and Cosmo didn't. Emily used past and future tenses—Cosmo didn't. And while both mixed actual words with other sounds, Cosmo did that much more than Emily.

In general, humans of Emily's age can learn the meaning of a new word almost instantly, whereas work with other parrots has shown that it takes them days or even weeks to take a new term onboard.

Taken together, these data suggest a dramatic superiority for Emily, but as the researchers note, she had unusual linguistic ability to start with. In fact, for the duration of the experiment, she produced about five times as much variety in her spoken language as a typical two-year-old would. Cosmo was right in there with those average two-year-olds. Not bad for a parrot![7]

Parrots

In another roughly similar 2024 study, the vocabularies of twenty-one African greys were compared with those of twenty-one children aged eight to eighteen months, the goal here being to expand the number of parrots involved in linguistic studies. The data was collected from parents and owners. This study showed that children use labels for objects, like toys, much more often that parrots do, even though parrots are fond of toys. They just don't discuss them, at least in the home context. By contrast, trained parrots, like Alex, use object names often.[8]

Children also used emotional expressions more, while parrots rely on greetings and multiword comments. The researchers point out that some of these differences are hard to explain. For instance, parents or caregivers are tuned to their children's emotions and so might notice them readily, while parrots might be reverting to their natural mode of expression for strong emotions: squawking.

The fact that parrots rely more on greetings could simply mean that they're left alone more often. But stringing together words is something the parrots in this study did much more often than the children, and to the scientists involved, many of these strings of words seemed to have been put together with a purpose, not randomly.

All this information would be extremely tempting to a pet owner who is fond of studying his/her pets, and parrots have the additional attraction of forming close relationships with their owners. As cool as the science is, it mostly lacks something that is usually there with parrots: humor.

In 2008, Liz Murray, a filmmaker in Vancouver, wrote, shot, and edited a short film called "More Than Just a Pretty Face."* The cast is a group of talkative parrots, and if you've never seen (or especially heard) them in action, it's a must-watch.

---

* You can watch Liz's film at https://vimeo.com/666567849.

Sorry, We're Out of Dogs and Cats, But . . .

Liz heard some great stories from parrot owners. In one, a friend of a parrot owner's boyfriend was sleeping over on the living room floor. When the house parrot woke up in the morning, he was overcome with curiosity, and walked over and pecked the friend's heel. The friend panicked, jumped up, and went running down the hallway with the parrot following close behind, calling out, "It's okay, it's okay!"

These are complicated pets. They are definitely not dogs or cats. Often joyful and amusing to be with, but demanding of company and potentially destructive if frustrated. As always, the human side of the equation determines whether the relationship works or not.

By the way, it's been reported that parrots learn new words best if they're spoken often, loudly, and with emotion. Hence cursing.

– CHAPTER 12 –

Ants

It would appear that socialism really works under some circumstances. Karl Marx just had the wrong species.
  —Bert Hölldobler and E. O. Wilson, *Journey to the Ants*

Horses and parrots, although outliers, are still well established as pets. Ants—not so much. But step back a bit, and think not just of a single tiny spindly-legged ant that might get squished underfoot, but of something much grander: a colony, assembled from—and by—thousands of those ants. Ants are social insects, and have evolved to build societies in which the insects, even though they are all the same species, adopt different roles and may even look radically different from each other.

The colony is an organism: Different ant "castes" have different jobs in the colony, but the overall goal is evolutionary success for all. This is where the colony gets interesting. Somehow, many ants spend their time and energy taking care of the queen's offspring. If evolution holds true for ants, as it surely does, there must be some evolutionary advantage to doing that. Act on your queen's behalf, and it works for you too—because of the genes you share with her.

But there are still many deep mysteries about how an ant colony flourishes, how the castes are maintained, how rebellion is suppressed, and what final threshold must be crossed for the queen to give up her only job: the egg-laying monopoly. It's complex—many surprises are still in store for scientists—and it's all done with smell.

Why would you bother having an ant colony? Really, it's not a bother—once the colony is set up, there's not much you have to do, certainly much less than caring for a mammal. But ease of ownership aside, is there a reason you'd want one? I think so. The reason isn't so much the usual emotional attachment to a pet; it's a glimpse into an alien life, a unique social system. Why *wouldn't* you want a close-up look at the day-to-day life of such an animal, which most of the time is hidden from view? And you can do it without leaving home!

If you are not an ant person, why not? Is it that they're *insects*, and social insects at that? In the early 1990s, one survey listed "aversion, anxiety, fear, avoidance, and ignorance" as the preeminent attitudes toward invertebrates like ants. Innate dislike, fear of disease, discomfort with the huge swarms of tiny things, and the "apparent lack of a sense of identity and consciousness" were cited for those negative feelings.[1] But ants are a group of animals that should be given the opportunity to help us expand our narrow human view of life on earth.

E. O. Wilson and Bert Hölldobler's Pulitzer Prize–winning book *The Ants* is 732 pages long and weighs 3.4 kilograms (7.5 pounds), dimensions that might lead you to believe it's hardly worthwhile to write more, but in the context of pets, not nearly as much has been written, least of all why you should own an ant colony. Playing on the idea of the "superorganism," the colony itself can be—should be—thought of as a single pet, an organism where the individual ants play the role of single cells, tissues, even organs.

If you were to have an ant colony, you would have representatives of one of the most numerous animals on earth. Ants are found everywhere in the world except Antarctica, Greenland, and some scattered islands across the world's oceans.

What about numbers? Surely one of the strangest things about estimates of the total number of ants in the world is that the first published number was in the children's book *Joan Embery's Collection of Amazing Animal Facts*![2] Although no studies were quoted, the number, one quadrillion (1,000,000,000,000,000), was pretty much in the ballpark of subsequent estimates in the scientific literature. In *The Ants*, Wilson and Hölldobler estimated there were ten thousand trillion (ten quadrillion) ants in the world; that's 10,000,000,000,000,000. They added that, despite the ant's minute size, that population probably weighed as much as all the humans on earth. The problem with these and other estimates was that they depended on guessing the total number of arthropods (animals including insects, crustaceans, millipedes, centipedes, and spiders), then estimating what percentage of all those are insects, and then just ants. With that number of steps, errors can magnify.

More recent measurements based on multiple sets of data derived from actual counts upped the estimated number of ants on earth to twenty quadrillion, double the number that Wilson and Hölldobler came up with, meaning that for every person on earth, there are more than 2,200,000 ants. These numbers are said to be on the conservative side.

The numbers alone say that these are animals worth paying attention to, but the diversity of species, habitats, and behaviors puts them over the top. An ant colony in a formicarium (an indoor ant habitat), as simple and uncomplicated as it is compared with life in the wild, still reveals the unusual, even bizarre qualities of ants that set them apart from most other animals.*

Two crucial features of ants are that their colonies are tightly knit societies and that stability is maintained by odor.

Dogs employ odor too, of course, using their sense of smell to perceive the world around them. But dogs can also see and have keen

---

* I'm describing ant life here as a composite of what's been found by studying many different species.

hearing, so while their sense of smell is extraordinary, it doesn't carry all the sensory weight. Some ants have decent but unremarkable vision, and many ant species don't really see much at all. While they do respond to some sounds, the auditory world doesn't play a major role in their life. But smell—it's everything. Without it, the ant colony would fall apart.

A colony is built around a queen that, once she is established, is an egg-laying machine. Most of those eggs hatch into female larvae; a much smaller number are males. Once a year, usually in summer, the queen alters her egg-laying pattern and starts producing males as well as a different sort of female. Both have wings and will participate in what's called the nuptial flight; on that dramatic day, they erupt from the colony, taking to the air with one thing in mind (not sure they have minds): mating. The air is filled with winged ants from different colonies.

Predators exult. A winged female released from the drudgery of the colony might mate with three or four males, but most females and males are either eaten or blown away by winds or, as E. O. Wilson put it, "splattered on windshields." All the males, whether they've successfully fertilized a female or not, die in a couple of days.

Most of the queen hopefuls fail to be fertilized or are killed before they can, but some succeed. They must reproduce: Colonies depend on having at least some of their not-yet-queens succeed to ensure the future. Dozens—even hundreds—of virgin females might launch on the same day.

A female that has beaten enormous odds and has been successfully fertilized in midair acts fast. She lands, breaks off her wings and may even consume them for their nutritional value (they're no longer necessary), and tries to find a place where she can dig a hole and create a shelter. She needs to start a brood before some ground-based predator gets to her.

Once settled, she begins to lay eggs. The newly crowned queen has stored the sperm she acquired during mating in an organ called a spermatheca. She'll keep them there, inactivated, for the rest of her life, activating and doling them out to fertilize eggs only as they're produced.* Which is exactly what she does as soon as possible, mobilizing some of those sperm to fertilize her first bunch of eggs.

These fertilized eggs hatch into newborn female larvae. (Any egg from which she withholds sperm automatically becomes a male.) The newly hatched females must make it through the larval and pupal stages to adulthood—that they're the future workers of the colony makes it a life-or-death situation. It's the responsibility of the new queen—by herself—to take care of these helpless newborns. So from the moment the virgin queen launched from the colony where she was born, mated in midair, and dug a new colony, she has worked nonstop.

But you know . . . patience. Once the queen has produced a working set of adult ants, her life settles down to a monotonous routine of feeding and laying eggs. In most ant species, she is enormous and mostly immobile. The rest of the colony is almost all female workers. They tend to the queen, the fertilized eggs, the newly hatched larvae, and the pupae. Some of them forage for food as the colony continues to grow. Some may specialize as soldiers/defenders. And then, again every year, special queens-in-waiting and males are generated from the queen's vast stores of eggs.

Colony expansion is de rigueur for ants. It's an evolutionary thing: spread the genes (of the queen) as far and wide as possible. But there are many interlocking steps that must be managed: The queen has to be the dominant egg-layer for most of the year, but at the right time, that control has to be ceded to the workers, among which will appear several potential queens and some males who prep for the nuptial dance. How does this work?

---

* Note that "the rest of her life" could be ten or even twenty years.

The social organization is based on chemical communication, and that, in turn, means odor, the second special quality of ant colonies. Smell and taste are the two human senses that enable our relatively primitive level of chemical communication. This makes the colony slightly alien: It's hard to imagine that a bewildering number of interactions—hundreds per second in a large ant colony—have almost nothing to do with sight or sound, the two channels we most depend on.

Imagine a chance meeting between two ants. Each immediately uses its antennae to touch or stroke the other to gather identifying chemical information. The information is contained within the waxy, greasy substance that coats an ant's exoskeleton. Embedded, dissolved, bonded, or somehow integral to it are an assortment of organic molecules that act as ID. Using the analogy of our sense of smell (which truly is nothing like the ant's), the castes of the colony all smell different. The queen, an infertile female worker, a fertile female, and a fertile male—if you met any one of them, in the dark, a dab of your antenna and you'd know.

But the chemical communication channel has much more depth than that. It is also essential to maintaining order in the colony. The queen alone lays eggs. All the other females are infertile workers. The routines of the workers bring them into contact with a variety of castes of ants, including the queen. When a worker ant is in intimate royal contact, the queen somehow (probably just by physical contact) loads queen-specific identifiers onto the worker's exoskeleton. They not only inhibit that worker from starting down the physiological path to becoming an egg-laying queen wannabe, but the worker also inadvertently spreads these queen chemicals to other workers during the repeated moments of recognition, ensuring that those workers will remain (infertile) workers too. Whatever the exact identity of these identity chemicals, they ensure the queen will continue to be the center of attention and the only egg-layer in the colony.

The evidence for this suppression of workers with queenly ambitions is clearer for the much-better studied honeybee. In the presence of the queen, only 0.01 percent of workers have activated ovaries, a mere 0.1 percent of these workers laid eggs, and only 2 percent of the eggs hatched.[3]

That all makes sense as long as the colony is booming: Workers are needed for work, and because there are larvae all over the place, it makes no sense to be producing potential queens. Every worker is needed to feed and take care of everyone. The queen's chemistry maintains her monopoly on reproduction.

However, as the season for nuptial flights approaches, the queen's veto power over the emergence of new queens is relaxed, or overridden by a small selection of workers that have the right body chemistry. There aren't many of them, but they do represent the future.[4] The details of how exactly this is accomplished need to be worked out, but it's a cinch that the smell of the individual ants is the medium of control.

When scientists dabbed ordinary worker ants with a synthetic version of a chemical displayed by workers changing to become fertile (en route to queenly status), they were instantly attacked by other workers—as long as there was already a queen in place. In the absence of a queen (demonstrating the need for one), these princesses were tolerated.[5]

It's remarkable how versatile the repertoire of signals is. Usually worker ants will destroy random eggs that were laid by a worker, but won't destroy queen's eggs—they smell different. In fact, the queen's egg stimulates workers to care for it; it also likely inhibits any tendency they might have to produce their own eggs.

These chemical labels also distinguish nestmates from uninvited strangers, which, once identified as foreign, will immediately be attacked. That in itself is pretty cool, but it's more subtle than it sounds—there's always a balance to establish between false negatives (letting in a foreigner by mistake) and false positives (excluding a nestmate).

A foraging ant will lay down a chemical trail to newly discovered food as it returns to the colony to alert the others. They touch antennae to ensure that the recruits have some familiarity with the food source.

It can get pretty wild: One species enters a colony chemically disguised as a queen, and the larvae hatched from her eggs are tended to by the host workers. Wherever there's a system, there will be cheaters.

I've barely scratched the surface of the life of ants—haven't said a word about the terrifying and all-consuming mass migrations of army ants or the exceptionally weird behavior called autothysis, where worker ants defending their colony against attack will bend their abdomen so far forward it ruptures, spraying a sticky mess all over any attackers in the way and dying in the process. But those phenomena would be very difficult to see in a pet colony, mostly because the species responsible wouldn't be available.

That is a reminder of one important caution about starting an ant colony: Be aware of the ant species that are native to where you live and house only those. In the unlikely case of a mass escape, you don't want to run the risk that non-native ants could set up shop in a new ecosystem.

It's a rare pet that demonstrates a whole different way of existence to us, and does it all in an enclosed space. You're unlikely to connect emotionally to your ant colony (although it's not that different from connecting emotionally to your favorite team, is it?), but keeping an eye on its goings-on gives you the chance to celebrate the one-hundred-million-year history of the ant. As we all should.

– CHAPTER 13 –

Hydras

Pets are our companions, playmates, protectors, fashion statements. Horses and parrots satisfy those criteria, albeit with some complications that the owners of dogs and cats avoid. As we've seen, though, they have qualities and history that enhance their differences from our most familiar pets. Ants demand attention but little else and, if watched carefully, reveal ways that life works you might never have imagined. They are a constant reminder that not all life operates on the same size scale as that of humans.

There are, of course, many other animals—birds, reptiles, amphibians, and insects—that could expand and enhance our narrow view of life on earth. But being who we are, we are fascinated by animals that have abilities we envy, and some of them become experimental lab animals because of it.

One fascinating example that has already achieved pet status is the axolotl, a salamander that never reaches the typical salamander version of adulthood. Most salamander species begin life as aquatic creatures with gills, but metamorphose into air-breathing land animals. The axolotl stops short of that, retains its gills, and lives underwater for its entire lifespan, which can be ten years or more.

Axolotls are in the strange situation of being almost, or perhaps actually, extinct in their natural habitat in Mexico, but proliferating as lab animals and pets in captivity. Why? They can regenerate lost body parts.

I have hesitations about them as pets. One is that virtual extinction in the wild. Purchasing an axolotl online or from a pet store requires verifiable assurances that you're not robbing from the struggling wild population. I'm also not totally comfortable about acquiring a pet whose uniqueness (besides, believe it or not, having a cute face) is that you could repeatedly amputate its limbs for your amusement. Scientists do it to continue learning about how it happens so that one day it might be possible to encourage the same sort of regrowth in humans. That seems worthwhile to me, but doing it for amusement seems different. That's why I've chosen a different, non-pet animal that provides all the weirdness of regrowing body parts combined with a few other delights.

In 1744, France declared war on Britain, the first rules of golf were written, and Abraham Trembley wrote his epic scientific work *Hydra and the Birth of Experimental Biology*. (I bought a copy of the translation years ago—it cost me $45. Today you can pay $300 online.) Trembley's book is a landmark in science writing: His inventive experiments, their details, his thought processes—you do feel like you're standing beside him as he works.[1]

But I'm not here to argue for Trembley's writing style so much as its content, because *Hydra* is an introduction to a creature that I'd argue should be a common pet.* Hydras are so fascinating to watch that you might wonder why they're not already in homes everywhere. They are barely visible yet are ambush predators! They have octopus-like tentacles—stinging tentacles at that! They can do somersaults (though even at their fastest they might travel about fifteen centimeters [six

---

* Named for the mythical, many-headed monster eventually killed by Hercules. That Hydra and these share the ability to regrow lost parts. If decapitated, the mythical Hydra grew two replacement heads, not one.

Hydras

inches], in a day)! And, most amazing of all, you can cut a hydra into pieces and each will develop into a new hydra!

The one barrier to their international fame? Size. Just as the bulk of the horse restricts where it can be kept, the hydra's *lack* of bulk can make them a challenge even to find. A large individual can have as much as a twenty-five-millimeter-long body (about an inch), but most are smaller, their bodies are slim, *and* they're not particularly colorful. And the neighborhood they live in, pond water, preferably overgrown with plants and green scum (algae to you), makes spotting them even more difficult for potential pet owners. (Funnily enough, if pet owners keep freshwater fish, they may have hydras already living in their fish tank.)

But this disadvantage is outweighed by the many advantages of the hydra as a pet: Collecting them from the wild would in no way endanger them—they're extremely plentiful; they are devout carnivores, for those who like to see their pets eat live prey (looking at you, lizard owners); they can reproduce by budding new hydras from the body of the old;* while cat and dog owners can enjoy behaviors of their pet that have developed over millennia of domestication, hydra owners get to explore a hidden world, an entire ecosystem in a mason jar; and, as I mentioned above, you can cut them into pieces and end up with more hydras than you started with. It's even possible to disassemble them into their individual cells, and each cell will regenerate a new, intact animal. What's not to love about that?

Hydras are simple structurally, complex behaviorally. They're not much more than a simple tube, with an opening at one end that takes in food and expels waste. Surrounding the mouth is a circle of long, extremely fine fibers—the tentacles—that, unlike with octopuses, have

---

* Budding is a neat way of coping with the fact that the cells in the hydra body are always dividing. Rather than continuing to grow to gigantic sizes, the parent animal creates buds that absorb most of the newly created cells.

no set number, at least partly because new tentacles are being generated all the time. Trembley spent much of his time examining how the tentacles enable the animal to capture prey, noting that if a tiny millipede or daphnia (the water flea) even so much as touches a tentacle, the unfortunate victim is almost always unable to escape. In fact, its frantic movements result in more tentacles wrapping it up and at the same time conveying it to the mouth. The hydra seemed to Trembley to have an extraordinary appetite, consuming virtually everything that it could catch, one morsel after the other, the normally slim body expanding to accommodate.

His groundbreaking discovery was the fact that he could slice hydras up and have each piece regenerate an intact organism. Ironically, he first did that (why *would* you do that?) because he suspected that hydras might be plants. His initial observations of the first species he studied, a small green variety that didn't at first appear to move and was shaped something like a plant, was the inspiration. He reasoned that if it were a plant, cutting it in two might yield two plants. It did, but as Trembley watched this amazing result, he was also making other observations that weakened his conviction that hydras were plants, and in the end, he realized they're animals, but you *can* cut them in half and each half will become a new animal.

Wondering how far he could take this, but realizing that hydras are too small to keep cutting them in half, he did it sequentially: cut one into four pieces, let them mature into complete hydras, cut each of those four into two or three pieces, rinse and repeat until he had fifty pieces from the original animal. Each flourished as an adult animal. Immortality in the local pond—or a fishbowl.

He didn't just cut them across, separating top from bottom, but also lengthwise, resulting in two half-hydras with their insides exposed from head to tail (like an open hot dog bun). After pausing for a while, the edges of each half would begin to curl toward each other until they joined—again, two adults resulted.

Hydras

Here's one of my favorite examples of Trembley's careful but ingenious experiments: He was wondering if the hydra had an ability to sense the proximity of prey, even if it hadn't wandered into direct contact with one of the tentacles. He used as his test animal one that was budding from a parent hydra but not yet separated (note Trembley's term for a bud was "polyp"): "Choosing a moment when the heads and arms of each were turned in opposite directions, I dropped a small worm on the arms of the young polyp. Instantly, the mother turned its head and set about seizing the worm."

Smart, but then Trembley went one dramatic step further: "I sliced off the mother's arms completely, cut off her head, and gave the worm back to the young polyp, assuming the prey could no longer be stolen from it."

Half an hour later, he returned and saw, to his shock, that the worm was halfway into the mother's stomach, not through her mouth (she didn't have one anymore) but "rather through the opening that was formed at the end of this headless stump by the turned-back edges of the anterior portion of the mutilated animal." (Remember, all would be well, as the mother would soon be whole once again.) So somehow, they're aware of the proximity of prey, but how? Trembley could find no sign of eyes, and indeed, hydras have no eyes, but it's likely that they can detect chemicals washing off prey and respond with a simple "head in the direction where the concentration is the highest" technique, the way many hunting animals do. You might have thought they'd need eyes to respond to light, as they do when in a jar by the window, but light-sensitive chemicals could be an adequate stand-in for image-forming eyes.

Remember, this was the early 1740s, and Trembley, as inventive as he was, lacked the kind of technology that would have allowed him to make even more dramatic discoveries. And that's where we are today.

One major advance is the understanding of how the tentacles work. Trembley was puzzled by them, noting that while they stuck firm to a prey animal, tangles of tentacles from several hydras could unwind themselves faster than you can untangle the charging cables in your desk drawer, suggesting to him that hydras have full control of the process. He would surely be astonished to know that studded along the length of every tentacle are hundreds of tiny capsules with coiled threads inside. When contact with prey is made, the threads are explosively released, like harpoons, penetrating the prey *and* releasing a neurotoxin. Trembley observed that even after a prey animal was contacted, it would continue to struggle, suggesting that the toxin is not completely effective, at least in small doses. Regardless, the release of the thread is incredibly fast: High-speed video has revealed that the firing of the "harpoon" takes only a few hundred billionths of a second, one of the fastest biological processes known.[2]

As much as I'd like to see hydras as pets, it looks like they're more likely to become a new and highly versatile lab animal. For instance, hydras have no brain, just a simple network of neurons, and that entire set of neurons has now been mapped; while that's been done for other simple organisms, it's been taken a step further with hydras by understanding how the few hundred neurons are wired to muscles to create every single behavior (elongating, somersaulting, contracting). So these tiny hunters are now to the point where they can serve as a model for better understanding how movement is coordinated.

But they are also on their way to becoming a key to understanding the biology of aging. The fact that a single hydra can regenerate new, fully formed animals after being sliced and diced suggested it has a never-ending supply of stem cells, those cells that have the potential to specialize into any kind of tissue. The front end of a half-hydra can create the back end and vice versa. It's now known that there are four different stem cell lineages in hydras, which together can re-create the entire body and, in the scientific terminology, are capable of "continuous self-renewal."

Perhaps the most amazing demonstration of this ability is that

Hydras

hydras can be repeatedly forced through the narrow opening of the fine tip of a pipette until the shearing forces have ripped the animals apart, leaving nothing more than single cells. If enough of those cells are then cultured together, including the right mix of cell types, these individual cells will collect and, in two or three days, form complete animals.

While axolotls demonstrate impressive feats of regeneration, they really are amateurs compared with hydras, but this is the price paid for complexity. In general, the more complex the body, the lesser the ability to regenerate lost limbs or, in this case, whole bodies. Axolotls can replace limbs, lizards their tails, and humans the tip of a finger.

I admit that classifying hydras as pets is a stretch at this point—they don't pay attention to us, they hardly seem like animals at all (remember Abraham Trembley at first thought they were plants), and they demand close-up viewing. But while you can't teach them to fetch, you can become attached to them, especially because—being immortal—they can become familiar. Friends even.

– PART IV –

People, Pets . . . and More People

– CHAPTER 14 –

Pet People

Dogs come when you call them; cats take a message and get back to you later.

A dog thinks: "This person's great. She feeds me, she pets me, and she plays with me. She must be a god!" A cat thinks: "This person's great. She feeds me, she pets me, and she plays with me. I must be a god!"

Are you a dog person or a cat person? Or any sort of pet person at all? Is this even a meaningful question? What if you'd owned cats all your life but suddenly were responsible for a dog that had lost its owners? Would you reject it or simply adjust? What if that replacement pet were an iguana? If you love animals, you love animals, right?

Many questions and unfortunately not as many answers as you'd like, but there are studies that have at least shed a dim light on the characteristics of pet owners. But piecing them together, trying to create some sort of unified picture, is tricky: Often the research uses different measures of personality, or is limited to certain kinds of people (like those who take their pet to the vet), and samples vary widely in size. Finding consistency across studies is challenging.

Incidentally, while there is a lot of research on the personalities of pet owners, I haven't found much on the personalities of those who would never own one and actually hate them. That is likely because it's easier to gather information from pet owners through veterinarians and pet supply stores, but regardless, even the little that's out there isn't helpful. For instance, searching online for "people who don't like dogs" yields many results, each of which goes no further than an explanation of what that person doesn't like about dogs: "They drool," "they want to lick your face," "they smell," "they're unpredictable," "they run to the fence and bark at you," and "their owners are insufferable." None of this tells us much about these people and their dislikes. Maybe it's something about their personalities, but if so, I haven't found it yet. One study showed that genetics plays a role in how much a pet owner plays with a pet, but that only goes a short distance toward explaining why some people just don't like them.

The situation is no different from efforts to understand owners of exotic pets. Personality is tricky and complex, some personality attributes are given different labels in different studies, and studies that ask pet-lovers or -haters to describe their own personalities are too dependent on the self-judgment accuracy of the participants. In general, when asked to describe their own personalities, people often err on the positive side. With that qualification in mind, efforts have been made to characterize pet owners.

Several, though by no means all, studies on the personalities of pet owners focus largely on what psychologists call the "Big Five" of personality, the five qualities that, taken together, are generally agreed to describe a person's personality adequately. No time or space here to delve into these in any detail, but here is a list of the five with a brief description of each:

Extraversion—sociable, warm, excitement-seeking
Neuroticism—anxious, easily upset, stressed
Openness—creative, imaginative, accepting of new ideas
Conscientiousness—cautious, responsible, focused
Agreeableness—kind, considerate

The degree to which anyone's personality contains elements of these is determined by having them answer standard sets of questions, and the result is a mix of the Big Five. These at least give you a sense of what psychologists look for when they try to define a "cat person" or a "dog person."

So, for instance, in 2010, Sam Gosling's team at the University of Texas found, out of 4,565 people recruited online, that dog owners rated higher on extraversion, agreeableness, and conscientiousness but lower on neuroticism and openness than cat owners. The study attracted participants from all over the world, ages from ten to ninety-five, 63 percent of whom identified as female. Separating males from females in the results (to detect any effects due to gender rather than pet preference) didn't appreciably change the results.

A few years later, another study expanded the criteria from the Big Five to a list called the 16PF in an effort to capture some subtleties that might be glossed over when using the fewer criteria of the Big Five. For instance, in this case, "extroversion" doesn't stand alone, but is divided into "warmth" and "social boldness," which do sound like very different characteristics. Someone who's low on warmth and high on social boldness might be someone you'd like to keep on the other side of the room, whereas low on boldness and high on warmth just sounds charming. Yet both those people could score identically on the umbrella term "extraversion."

In this study, six hundred undergraduate psychology students answered questions like "Did you have a pet growing up?" If yes, then "Check all that apply: dog, cat, rabbit, bird, fish, ferret, tortoise, snake, horse, pig, lizard, turtle, duck, frog, hamster/gerbil, other." There were also questions about pet preferences and personality; then the massive amount of data was sliced and diced statistically to generate these results: "These findings describe the personalities of the average cat person as shy, solitary, impersonal, serious, and nonconformist, but also creative, sentimental, independent, and self-sufficient." You know what, cat people? It could be a lot worse, although that subset of shy, solitary, impersonal, serious, independent, and self-sufficient doesn't sound like life is a party.

Dog people? "Grounded, pragmatic, and dutiful, as well as warm, outgoing, sociable, expressive, and group oriented." Life *is* a party! However, as a dog owner, I am duty-bound to add that the study confirmed that cat people scored higher on general intelligence. The authors also pointed out—yet again—that despite the differences, overall the personality profiles of cat people and creative people are remarkably similar.

Gender differences lent some interesting touches to the data. Emotional sensitivity was significant only for cat-loving men, not women, and reasoning (see general intelligence above) significant only for women who preferred cats.

Some of these findings point to people choosing appropriate pets. For instance, males were found to be livelier than females, and males tended to prefer dogs. The simple act of dog walking fits that scenario. The shy, solitary, impersonal cat owner also has chosen the right pet.

But are we consistent? I lived a good part of my life owning lizards and turtles, then only cats. If I had participated in a study during those years, I would not have been included among the dog owners. But now I would be, even though my personality has likely been pretty much consistent over that time. Could it really be argued that having a dog changed my personality? It's not impossible, but also impossible to know. In an article in *Psychology Today*, Susan Whitbourne also pointed out that 43 percent of the people in the survey were neither exclusively pro-cat nor pro-dog.* They either liked both or neither, and yet they had strong personality parallels to those with pet preferences.[1]

---

* People who like both cats and dogs? I owe it to the *Guardian* columnist Zoe Williams for the terms "ambipetxdrous" and "bipetsual" for those of us, myself included, who like both dogs and cats.

But there's much more data. Returning to the Big Five of personality, another study examined and detailed those characteristics in cat owners. Owners who scored higher on neuroticism were less likely to let their cats outside and reported their cats were more aggressive or fearful. On the other hand, the more extroverted the owner, the more likely the cat was allowed outside, and conscientious owners claimed their cats were less aggressive, anxious, and gregarious.

One curious finding: Owners scoring high on neuroticism tended to report their cats were overweight; agreeable owners, the reverse. The authors point out that these findings parallel to some degree what's already known about parent-child relationships.[2]

I'll add a little more to this mix with some findings by Stan Coren at the University of British Columbia. Having done his own survey, which produced findings not unlike those above, he had some additional comments.

He asked both cat and dog owners if they'd cross over and accept a gift puppy or kitten, providing they had space and no objections from the people in their life. "More than two-thirds of the cat owners (68 percent) said that they would not accept a dog as a pet, while almost the same number of dog owners (70 percent) said that they *would* admit the cat into their household."[3]

I'm tempted to read into this something about the differences in the pet owners' personalities, but the first reaction I got when I put this question to pet-owning friends was "Oh yeah, cats are less work." The second reaction was "Yes, I'd take the cat, but only because it was 'forced' upon me," illustrating that yes, personality might play a role, but maybe in unexpected ways.

There are many more studies, but it's worth asking, aside from confirming what many pet owners always believe (that there are differences between dog and cat owners), can this information in any way be useful? Maybe.

Every year, millions of pet owners give up their pets. Some find another home with a family member or friend, but many end up in shelters. Of those, significant numbers are euthanized. Current numbers for the United States are more than six million animals entering shelters annually, with just shy of a million being euthanized. The reasons for giving up a pet include problem behaviors like aggression, health issues, and size, but human issues are also crucial, including personal housing, veterinary costs, allergies, and abuse. While close to a million dogs and cats are euthanized in shelters, you can easily find estimates that ten times as many are killed by abuse (but so far I haven't been able to find an original source for that number).

These data may point to a practical application of owner personality, and that is, should personality assessment be taken into account when a human chooses a pet? It's already well known that if a child lived with a pet when growing up, he/she is likelier to have one when an adult. But could that predisposition be fine-tuned?

And what about the other side of the relationship coin—the pet's personality? "She is flamboyant and sparkly, a real mover. . . . At times she dances instead of walks, kicks up her heels and jumps, twists her head and shoulders sideways . . . and yet what a genius she is. Lively, curious and exuberant both in social situations and solitary activities such as moving about and fishing." That's biologist Bob Fagen speaking. And no, he's not talking a daughter, a friend, or even a pet. He's talking about a bear, a bear living in the wild code-named "C," one of a group he studied in Alaska for years.[4]

Fagen's studies of behavior show that animals have what you can only call personality, and different bears have different personalities. That is an intro to the question: Could dogs and cats—and other pets—also have personality characteristics not unlike our own, and should the choice of a pet be based more on personality than cuteness, maybe even to the point of ensuring the human and the pet are a good match? It's common for vets to caution people against choosing obviously poor matches, like a high-activity dog for a sedentary person, but is there enough science to provide more detail?

Sam Gosling, whom I mentioned above, took a shot at applying the Big Five personality assessment to a range of animals, albeit combined with other psychological measures. The results confirmed that a wide variety of mammals, from primates to hyenas and donkeys, vary reliably when it comes to measures like extraversion, neuroticism, and agreeableness. Even octopuses qualify—the octopus that sticks to its den or conceals itself behind a cloud of ink (low in extraversion) could be compared to the human who stays home on a Saturday night.

That study was published in 1999, and even Gosling admitted it was tentative, likening it to a first sketch. He and his team then designed a study to compare dog personalities directly with those of their human owners. First, owners assessed their own personalities, then their dog's. Armed with those two personality profiles, the experimenters then recruited people who knew both the owner and the dog and had them corroborate the personality findings, and finally had an independent person watch the dog's behavior in a park and compared the resulting assessment with the dog's personality ratings to ensure the two were consistent. And they were, putting the idea of direct comparisons of human and dog personalities on a much firmer footing.

Then, in 2013, a team at Oklahoma State University concluded that four characteristics were common to both humans and their dogs: sharing possessions, love of the outdoors, the ability to get along with others, and a tendency to destroy things (or not). This last category piqued my interest. I'm familiar with dogs ripping toys apart, but owners' destructive behavior? As the researchers stated, "It would be interesting to research this further, to determine if people select dogs that are more destructive or it is people's treatment of their dogs that causes them to become destructive."[5]

In 2023, a review of everything published on the match between owners' and dogs' personalities largely backed up the research I've so far mentioned. The researchers in this case added "attachment" as a second important factor, arguing that the attachment between a dog and its owner is similar to the attachment between human caregivers and infants (wanting to be near the caregiver) and is facilitated by the now well-known effect that a dog gazing into its owner's eyes increases levels of oxytocin in both.[6]

They then gave voice to the idea that, while factors such as personality and attachment create good bonds between dogs (and likely other pets), these are seldom the motivation for acquiring the dog in the first place. Cuteness and good looks generally prevail. Given the large numbers of pets that are eventually given up, the authors wonder if incorporating some way of assessing whether a human and a prospective pet are a good match, based on personality and attachment, would not be a much better way. The public is so far generally unaware of this research, and we are pretty far away from being able to put a suggestion like this into use, but if there were a practical way of doing this, it's a good bet the results would be worth it.

In the meantime, choices are made with little regard for suitability and the longevity of the relationship. Not surprising really—thousands of years associating with pets has resulted in a bewildering array of human attitudes and behaviors toward them. Nowhere is this clearer than in the naming of cats and dogs.

– CHAPTER 15 –

Pet Names

Dog and cat names unfortunately dominate the literature on pet names—it would be lovely to know if there have been trends in naming budgies or ferrets. However, there are still curious features to unearth, one being that a historical look suggests that pet names and the inspiration behind them have changed dramatically over the last few centuries. Names continue to evolve today, although relatively slowly. Another apparent feature is that while names formerly addressed some characteristic of the pet, today they're driven more by what the owners are paying attention to, especially in pop culture.

A scan of medieval texts came up with these dog names: Sturdy, Whitefoot, Hardy, Jakke, Bo, Terri, Troy, Nosewise, Amiable, Nameles, Clenche, Bragge, Ringwood, and Holdfast. Not sure you'd hear many of these, although Jakke, Bo, and Troy could probably be in the local dog park. But Nosewise? Clenche?[1] My favorites among dog names are Megastomos ("big mouth"), Fortuna, and Mopsulus. As far as cats go, "Gilbert" was common, but not so much as a specific name for a specific cat, but rather as a generic term for a tomcat in England, often shortened to "Gyb."

Medieval bird names tended to be pretty direct too: Birds like magpies were often called "Mag," robins "Robin," and sparrows "Philip" (okay, I realize that isn't obviously straightforward). A famous poem written in the early 1500s by English poet John Skelton is an ode to a sparrow named Philip after he was killed by the cat named Gyb. The poet and satirist Alexander Pope (1688–1744) had a series of Great Danes, all named Bounce.

William Safire, writing in the *New York Times* in 1985, revealed some historical trends in dog naming. He began by arguing that "in olden times" (while not saying exactly when that was), dogs were named according to their disposition—Friskie, Rover—or for their appearance—Spot, Rags, and Blackie. Straightforward, although Safire adds that sometimes the straightforward could be sarcastic: A dog that spent most of its time resting might be called "Lightning." Royalty played a role too: King, Prince, Duke, Queenie, Princess, and Rex.

Safire argues that by the mid-1980s, dog owners had turned to using human names, perhaps because they had never applied them to their children, so names like George, Daisy, or Charlie became common. (Fair enough—there are now dogs in my extended family named Enzo and Freddie.) Followers of Safire's *New York Times* column On Language, whom he called the "Lexicographic Irregulars," responded to his request and forwarded stacks of pet names, mostly dogs (12,000 in all). He noted that "the biggest pile is 'Names of People,' and the names that appear most frequently in it are Max, Belle, Ginger, Walter and Sam. (Sam is a dog; Samantha is a cat.)"[2]

That was 1985—is it any different today? Embark, a dog DNA analysis company, suggests these were the top dog names for 2024: Luna, Bella, Charlie, Lucy, Daisy, Cooper, Bailey, Max, Sadie, and Penny. (What, no Ringwood?) On its website, Embark goes to some lengths to explain the process it used to create a list that is representative of the two million dogs whose DNA it has analyzed.[3] Two million is a large sample, but let's remember that there are likely about seventy million dogs in the United States that live in homes and therefore are likely to have a name, and people who have their dog's DNA analyzed *must* be special in some way, so we can't take this as the final word.

Camp Bow Wow (yes!) lists these top ten: Luna, Charlie, Bella,

Pet Names

Lucy, Cooper, Daisy, Max, Milo, Bailey, and Sadie—nine names the same as those from Embark! Trupanion Pet Insurance also lists nine of the same names. Unless they're colluding, those dog names seem to be at the top of the list.

Camp Bow Wow's animal health and behavioral expert Erin Askeland points out the huge influence of pop culture: "Maverick had the highest year-over-year increase, and Goose came in third, names that could be paying homage to the 2022 blockbuster *Top Gun: Maverick*. Poppy saw the second largest year-over-year increase, likely inspired by Anna Kendrick's character, Poppy, in the 2023 film *Trolls Band Together*. Gigi and Millie also increased year over year, possibly attributed to model Gigi Hadid and actress Millie Bobby Brown. We expect to see an influx of names like Loki, Greta, Rocket, and Barbie in the coming years."[4]

Switching to cats, the Scottish newspaper *The Scotsman* reports a top ten including Luna, Bella, Milo, Poppy, Coco, Nala, Willow, Loki, Mollie, and Charlie. When you have four top dog names identical to four top cat names (Charlie, Luna, Bella, and Milo), I sense a boring sort of consistency, even lack of imagination.

An Australian survey (petsecure.com.au) listed the top names for 2014, and about 80 percent of the same names were present. Only if you go back more than two decades do you begin to see some differences. For instance, Pet of the Day listed the top ten dog and cat names in the UK for 2003. Top ten cats: Tiger, Charlie, Sooty, Jasper, Tom, Lucky, Smudge, Poppy, Gizmo, and Misty. Top ten dog names: Charlie, Sam, Holly, Jack, Cassie, Ben, Benjie, Ollie, Max, and Coco.

So it took decades, but there's been significant turnover since 2003. That suggests that the pop-culture-inspired names, like Maverick, don't last long, even though one survey showed that a third of pet owners used names from movies, TV, or video games. Although, to be honest, some 1990s pop-culture-inspired dog names are coming back; Rover.com claims that Alanis is up by 547 percent. Ginger Spice is rising too, as is Spice Girl.[5]

But why, oh, why? Where does the inspiration for names like these come from? There is, believe me, a shortage of scientific research on this. However, psychologist Becky Spelman, founder of the Private Therapy Clinic in the UK, outlined some of the reasons she thought were behind dog names, and no surprise, the owners were primarily thinking about themselves. Names were chosen, according to Dr. Spelman, to allow owners to showcase their creativity, interests, or sense of humor; to demonstrate exclusivity and intimacy in the relationship; to serve as conversation starters; or to set the pet apart from others.[6]

There's also an argument that the preponderance of human names (like Max, Bailey, Daisy, Charlie, and Cooper) in the most popular lists represents a growing attachment to our pets, in effect, seeing them as more human. One observation supporting this is that each year there's a significant crossover between baby and pet names. There's a temptation to link the predilection to give a pet a human name to the fact that 97 percent of pet owners in the United States say their pets are part of the family; 25 percent would change jobs to spend more time at home with a pet, 23 percent would move, 18 percent would get a new roommate, and 16 percent would break up a relationship—with another human.[7] It's possible that these numbers are higher than they used to be, but surveys like this were not common in the past. It's also difficult to combine these two data sets and argue there's a causal relationship between them.

However, a study published in 1987 showed that if pets were seen as a member of the family, they would be more likely to get a human name, and even more likely if they were considered to be "extremely important or very important members of the household." That these conclusions were drawn from the same survey of the same pet owners suggests they are indeed cause and effect. Incidentally, the tendency to bestow human names on cats was slightly higher than on dogs.[8]

And what do the pets think? Not much, presumably, at least to the nuances of the names, but the auditory qualities are relevant. Shorter names are said to be better; for instance, a two-syllable word with each syllable a different pitch is supposed to be easier for a dog to discern than polysyllabic words. But both cats and dogs recognize their names, unless, of course, they're either ridiculously long or can easily be confused with other words: Yelling "Axe the Tax" would be a nightmare for your dog Max. If you ever face such confusion, take heart from the fact that you're not calling an American Kennel Club champion, which might have a name like GCHP3 DC Leoralees Lets Boogie with Barstool Mw, GCHP3 CH Fenice Le Beaudreaux, or GCHP2 CH Cinnibon's Bedrock Bombshell.

Pet names have little influence on the pets themselves, but those Kennel Club champions just named represent a much more profound human manipulation of pets: the invention of new dog breeds.

– CHAPTER 16 –

The Whimsy of Dog Breeds

The Fédération Cynologique Internationale, the International Canine Federation, recognizes 356 breeds of dog. The American Kennel Club approves only 201 (these numbers continue to rise). If you're a dog breeder, those differences might be important, but for the rest of us, they're just a measure of how the two organizations set different standards for the inclusion of a purebred dog to the list. These include how many generations have been bred, how popular it is, how many there are, and how often the breed has competed in dog shows. Yes, it's a world apart from the other hundreds of millions of dogs on the earth. Still, new breeds are constantly being added: From 2000 until now, the AKC has listed fifty-six new breeds, an indicator of the unceasing desire to qualify new dog breeds that has been going on for the last two hundred years.[1]

Although new breeds are added every year, there are some ancient ones, including greyhounds, basenjis, salukis, and the Greenland sled dog. Dogs that look very much like greyhounds are depicted on six-thousand-year-old Egyptian paintings, but there's controversy over whether these could actually be basenjis or salukis. Greyhounds are also mentioned in some translations of the Bible (Proverbs 30:29–31):

There be three things which go well, yea, four are comely in going:
A lion which is strongest among beasts, and turneth not away for any;

The Whimsy of Dog Breeds

A greyhound; a male goat also; and a king, against whom there is no rising up.

Although that is clearly not a scientific reference. Basenjis appear to be the breed on cave paintings in Libya, and salukis in ancient Egypt. The *Guinness Book of World Records* cites nine-thousand-year-old cave paintings that portray saluki-like dogs, but obviously, visual resemblance doesn't make the strongest case for genetic evidence. That's why recent findings for the Greenland sled dog are more significant. An archaeological dig on the high Arctic Russian island of Zhokhov showed that people there 9,500 years ago were traveling great distances by dogsled. Genomic analysis of a dog jawbone at the site revealed it to be related to the modern Greenland sled dog.[2]

However, ancient breeds represent a minuscule fraction of the dog breeds today. Breeding thousands of years ago was intended to emphasize desirable functional traits (greyhounds were sight hounds for hunters; the Zhokhov dog pulled sleds), and that focus held steady through the centuries. For instance, the top five breeds of the 1880s were all working gundog breeds, but these pointers and setters had first been bred centuries before.

Working dogs are still bred today, but in the last hundred years, there's been a different trend: More and more breeds have been inspired largely by fashion. No treatment of a pet could better illustrate the adage "It's not the pet—it's the human."

Fashion is partly about appearance, although it's wide-ranging in the sense that it can signal anything from elegance to threat. But fashion can also mean an idea, a meme, that makes a sudden appearance, gains millions of followers, then fades. These too affect the breeding of dogs.*

---

* Of course this applies to more than just dogs. The Oscar-winning movie *Everything Everywhere All at Once* brought back pet rocks, offering "signature" versions of pet rocks for purchase on its website.

In 2016, the founder of the magazine *Lone Wolf*, Natalia Borecka, published an article called "100 Years of Canine Couture: Decade Defining Dog Breeds Through the Lens of Fashion."[3] Borecka portrayed some of the dogs that came and went as fashion accessories through the twentieth and into the twenty-first century: the Boston terrier from 1910 to 1930; then the spaniel, especially the cocker spaniel (most popular dog for sixteen straight years in the USA), and the standard poodle from 1960 to 1982, one of the longest reigns as most popular dog, only to be beaten out by the Labrador retriever from 1991 to 2022.

The twenty-first century saw the rise of smaller dogs like the Chihuahua, whose popularity was given a nudge by Paris Hilton and Britney Spears wearing them "like designer purses."[4] Then came the pug, followed by today's current number one, the French bulldog.

Some others took a run at being the most popular over the last century, but against this background of waxing and waning popularities (many of the dogs listed above are still in the top twenty), there have been startling flurries of interest in other breeds, prompted by movies.

There's a long history of dogs in movies. Teddy the Wonder Dog, a Great Dane who appeared in something like sixty movies from 1916 to 1924, most of them comedies produced by Mack Sennett, was the first major dog movie star, but it wasn't until the mid-twenties that the German shepherd Rin-Tin-Tin set the standard for courageous dogs that saved their human's skin again and again. Lassie continued the theme of the canine hero after the Second World War; Balto in 1995 and Chinook, the star of *Hero Dog: The Journey Home*, are the modern versions.

The anthrozoologist Harold Herzog and several of his colleagues have detected some startling responses to popular movies. The Disney movie *One Hundred and One Dalmatians* was re-released in 1985, and over the next eight years, the number of Dalmatian pups registered annually with the American Kennel Club rose from 8,170 to 42,816.

The Whimsy of Dog Breeds

Same effect with another Disney classic, *The Shaggy Dog*, in 1959: Registrations of Old English sheepdogs went up by a factor of one hundred over the next fourteen years.[5]

Overall, success usually has an immediate impact: The ten most influential dog films from 1926 to 1995 each generated eight hundred thousand more puppies registered in the decade following the film's release. However, what goes up with dogs (and baby names) must come down. Yes, Dalmatian registrations reached their peak in 1993, but then fell by an astounding 97 percent over the next ten years.

Sarah Weir and Sharon Kessler of the University of Stirling in Scotland dug a little deeper into the data by asking whether the way a dog was portrayed cinematically influenced the registrations that followed.[6] What stood out about their findings is that only one type of movie, that which stars a "hero" dog (one whose role is mostly to save and protect their humans), raised registrations. All others—including anthropomorphized dogs that could understand speech (or even, in some animated movies, speak), dogs situated in traditional suburban three-bedroom home settings, and dogs that came in from the wild—had no effect or even depressed registrations after the movies' release.*

This does raise the question: "Don't people take anything into account other than the dog having become fashionable or appeared in a recent movie?" Sometimes they don't.

The same researchers who originally tracked the link between a dog's movie appearance and its subsequent popularity asked the question: Do prospective dog owners rank their choices based on health, longevity, and desirable behaviors first? Not as much as you'd guess.

---

* The rabid Saint Bernard star of the 1983 horror movie *Cujo*, against all expectations, triggered a significant increase in Saint Bernard registrations in subsequent years. I'm sure you'd also like to know that Cujo was played by four real Saint Bernards, some mechanical ones, a black Labrador–Great Dane mix in a Saint Bernard costume, and a stuntman, also in costume.

Choices of purebred dogs are determined as much or more by fashion than any of these qualities that you'd think would be more important.[7]

The French bulldog is a good example. It is now the number one ranked dog by the American Kennel Association, but its history has been determined by social, not veterinary forces. The breed arrived in France during the Industrial Revolution, imported by lace workers who were losing their jobs in England.

The *bouledogue français* was immediately embraced by a swath of French society, from social elites to prostitutes, and it had an equivalent impact in America when it arrived in the late 1800s. Its popularity today is related to several factors: They're playful, small dogs that are more easily managed in cities, and they've enjoyed some pop culture hits as well: two of Lady Gaga's were stolen in 2021 (but returned); and James Cameron's *Titanic* featured a French bulldog named Gamin de Pycombe (based on a dog of that name that was on board the actual ship). However, I'm not aware of any movie-related bump in registrations, either from *Titanic* or any other film. (Does *Slightly Single in LA* count?)

But there are serious health issues with Frenchies. They are one of several breeds defined by the technical term "brachycephalic," meaning they have been bred over time to have shortened snouts. Other such breeds include pugs, English bulldogs, boxers, Cavalier King Charles spaniels, Boston terriers, Pekingese, and Lhasa Apsos. The motivation for this breeding trend is at least partly to make them more baby-faced and therefore cuter, à la the Konrad Lorenz prescription for "cuteness" I mentioned in chapter 5. But this preference carries with it a significant health downside.

The shortened snout crowds the structures of the airway and pushes them back down the throat, commonly causing something called brachycephalic airway syndrome, which compromises breathing in multiple ways: Nostrils are narrowed, and both the turbinate bones and the soft palate extend farther back in the airway, cutting down the flow of air. These characteristics—and there are more—create a domino effect that can seriously compromise breathing and general health. Dogs with a mild case breathe noisily; dogs with more severe cases tire easily and may even faint after exercise. The further the negative impacts extend into the body, the more pronounced the syndrome.

The American Veterinary Medical Association points out that brachycephalic dogs are prone to other medical issues including pneumonia and a variety of conditions affecting eyes and skin. French bulldogs, because they're heavier in front, can't swim, and owners are counseled never to leave them alone near open water. Even the idea that flatter faces make these dogs "cuter," in the Konrad Lorenz sense of looking more baby-like, and therefore more attractive, may be a handicap for communicating effectively with humans. A recent study showed that people trying to read the mood of such dogs relied much more on the dog's body language than its face, as if the facial expression, as cute as it might be, was no longer reliable or even readable.[8]

Some of these are genetic issues, unrelated to the brachycephaly, caused by inbreeding over long periods of time, but of course the point of the inbreeding has been to maintain the look of the breed. We are very familiar with the effects of inbreeding: It results in the accumulation of deleterious mutations, hence the incest taboo. For humans anyway. No such taboo for dogs. An article in the *New York Times* put the issue starkly. Depending on the breed, some are mated with the equivalent of cousins once removed, some with genetically equivalent half siblings or even siblings, and "some dogs are inbred to the point that it's as if siblings continued mating for multiple generations."[9]

The results are dogs like the French bulldog. Regardless of the initial motivations for its look, maintaining it is essential for its standing with the American Kennel Club. The club is particular about the look of the breed.

The Kennel Club recipe for the perfect *bouledogue français* is exacting and all-inclusive: The ears must be the proper shape and size, erect and situated in a precise position on the head. Even the texture of the fur of the ear must be proper. Such dictates also apply to the shape of the muzzle and the contours of the skull. The nose must be black. And that's just the head. The coat must be "brilliant" (and white, cream, or fawn, or some combination of these), the body "muscular."

Why do dog-lovers persist in acquiring dogs like the French bulldog, which are prone to an unusually large number of medical issues? One study of owners of four different breeds, including French bulldogs, found that the well-publicized health issues of the dogs they owned had not only not dissuaded them but may even have encouraged their desire to "take care" of a dog. They also were attracted to the dog's unique appearance, part of which is certainly connected to the cuteness factor. It was true, however, that French bulldogs' owners, at least in this study, were somewhat disinclined to buy another one.

Summing up all this information is difficult. Obviously, people buy purebred dogs for a variety of reasons, and on balance, it appears that social factors like fashion or movie appearances can be more important than health. This just underlines that what many people care most about when acquiring a dog is themselves. This is not a surprise.

While individuals seem happy to buy dogs that can end up living difficult (and expensive) lives, some governments and veterinary organizations are not so inclined.[10] A few years ago, the Dutch government reinforced a ban on breeding short-snouted dogs. Norway has banned the breeding of Cavalier King Charles spaniels, and for years, the British Veterinary Association has been urging customers not to buy brachycephalic dogs.

While it's encouraging to see the rise in sales of mixed-breed and rescue dogs, whose somewhat unrestricted mating limits the genetic damage caused by inbreeding, the human desire to be fulfilled by the purebred dog accompanying them is unquenchable. We won't be seeing it end any time soon.

I can't leave this subject without drawing a contrast between two breeds of dog, one that remains enormously popular, the other now extinct. They dramatize, like no other pairing, how dog breeds mirror human wants and needs as much as an animal could.

In early 2024, the Poodle Room, a members-only space at the Fontainebleau resort in Las Vegas, was launched. Predictably over-the-top, it featured, besides movie stars and caviar pillows, a white show poodle named Josephine. She's the mascot of the Fontainebleau, frequents Instagram, and occasionally is present to greet special guests, of which there were many this day.

But there is more than one "Josephine"; actually, there are several, including one called Patrón, who is actually male. When you're wearing a sparkly collar, though, no one really notices. Javier Torres, Patrón's owner, added, "He is a very poodle-y poodle," presumably because, properly fluffed up, he looks pretty feminine.*

The fancy poodle is such an outdated image (I had a wonderful standard poodle named Buster who was at his most joyful running in the woods looking for trouble), but at this party, all the Josephines served their purpose.[11]

Contrast, say, Patrón's life with any of the innumerable, but unnamed, members of the turnspit breed.†

Haven't heard of it? No wonder: They lived their pathetic downtrodden lives from the seventeenth through the nineteenth centuries. Turnspits have been described as rugged dachshunds, same short legs

---

* Ironically, the much-mocked "poodle-y poodle" cut—the one with the enormous lion-like mane and little poufs of fur here and there—has its roots in the dog's original role. Poodles were hunters' companions and had to swim to retrieve downed waterfowl. Fur was left long in some places (chest and joints) for warmth, but cut short in others to enhance speed through the water.

† Unnamed but not unnoticed: Both Shakespeare and Darwin mention them. And actually, we do know the name of one turnspit: Whiskey. He might have been one of the last of his breed, was taxidermied when he died, and is now on display at the Abergavenny Museum in Wales.

but a thicker body. They did not, like Patrón and Josephine, fulfill their role simply by looking good. Nor did they receive the adulation and care those poodles enjoy. They worked hard, and if they didn't work hard enough, they were punished.

Situated at the other end of the socioeconomic scale from the Poodle Room at the Fontainebleau, the turnspit was forced to run inside a wheel like a hamster. But unlike the hamster, whose physical challenge is to spin the wheel with no weight other than its own body, the turnspit ran a wheel that rotated a spit laden with a huge chunk of meat in an oven. The mass of the meat, added to the poorly lubricated connections between the turnspit's wheel and the oven spit, made for torture. Of course roasts of meat intended for families may have required several hours to cook, and the spit had to be kept turning. Whether true or not, consistent with their standing, it's said that if they faltered on the wheel, a hot coal might be tossed into the wheel to inspire greater efforts.

You'd think that praise might be heaped on the turnspit by the grateful and well-fed humans, but likely not: Families with any amount of wealth at all would hire a human to work the spit. Only the poorest, who had little time to cuddle or entertain the turnspit, would own the dog to begin with. It wasn't a dog; it was a mechanism.

A widely quoted comment from author Jan Bondeson sums up the turnspit's meager existence: "It became a stigma of poverty to have a turnspit dog. They were ugly little dogs with a quite morose disposition, so nobody wanted to keep them as pets."[12]

Whether adornment or tool, poodle or turnspit, purebred dogs exist solely at our pleasure; which lends a measure of respect to the hundreds of millions of mixed-breed, feral, or street dogs around the world—call them what you will—whose genetics are their own business.

– CHAPTER 17 –

Exotic Pets—Exotic People?

Most of the research on pet owners' personalities and how those turn breeds and pet names into memes is inevitably focused on dogs and cats. Even so, as we've already seen, drawing a clear and unambiguous psychological profile of those pet owners isn't easy. It seems just as challenging—if not more—to analyze the personality types of exotic pet owners. One study compared cat- and dog-lovers with those humans who prefer exotic pets, revealing differences when it came to the personality characteristic of openness: Females who owned dogs or cats were less open to new experience than females who owned cold-blooded exotics. In fact, the latter females were even higher on the openness scale than their male counterparts. Not surprising that they were comfortable with a "different" sort of pet.[1]

Because exotic pets are so much rarer than dogs and cats, it is more difficult to find owners who can be included in such studies and quizzed about their pet choices, which is frustrating, because you might suspect the rationales for those choices would be diverse and interesting—after all, they opt for uncommon, off-the-beaten-path companion animals.

Pairing humans with pets by personality is challenging in so many ways. Personalities, despite their categories and definitions, are complex and diverse enough to push back against being categorized and defined. Are there people who acquire animals for reasons other than

People, Pets . . . and More People

the usually expected desire for an animal companion? Yes, but how much does that say about those people? Are there types that stray from the norm?

Let's make a testable assumption: There are personality traits that correlate with owning exotic pets (although they may be more elusive than you might think). One of my favorite terms in this research is "personal identity projects," pets chosen not primarily as companions but instead as enhancements to the human's fashion or social status or even targets for dominance and control. Display becomes the most important feature of the pet, with the owner counting on reaping the benefits, whatever they are.[2]

That sounds like a dubious reason for choosing a pet! What if the fashion trend or emotional need (or whatever) passes, as it always does? We've already seen that with dogs, but what about exotic pets, the "strikingly, excitingly, or mysteriously different or unusual"?[3] A typical selection of exotics might include snakes (especially large ones), tarantulas, hedgehogs, ferrets, tortoises, iguanas, geckos, even giant African land snails and centipedes. Some of these might seem unusual but not dramatically different from a dog, cat, or bird. Hedgehogs can be cute, iguanas curious, and tarantulas mysterious. But ball pythons? Tigers?

If your reaction to this is that it's none of anyone's business who wants to own what pet, it's worth noting that exotic species are connected to the world's wildlife in two crucial ways: One, they are often taken from nature and sold into pethood, adding to the already crushing pressure on the world's endangered species. But it can get worse: In a weird turnaround, if released into the wild, some are able to establish themselves in novel environments—novel at least for them—and become invasive species, wreaking havoc on the local biosphere. So, just as it's accepted as helpful to know more about how likely a pet owner is to succeed with a dog or cat, it's worth knowing who wants to own an exotic pet, and why.

But it's complicated. Drilling down, you might suspect that people who love attention might choose an unusual pet as one of those "personal identity projects." Unusual attracts attention, both to pet and owner. The story of Narcissus, and the psychological profile linked to his name, is a big part of this.

Narcissus was the young man in Greek mythology who, having glimpsed his reflection in the still waters of a pond, became so obsessed with his own beauty that he spiraled into depression and died. The modern narcissist is also unusually interested in him- or herself, but the expression of that is not straightforward. "Narcissism seems to be related to contradictory processes and consequences: Narcissists' charisma and self-assuredness can give them tremendous energy that fascinates others, yet their aggressiveness and lack of empathy hinder their progress and turn many people off."[4]

Digging a little deeper, narcissism, usually characterized as pride, egotism, and a healthy focus on oneself, comes in more than one flavor. Extremes include "vulnerable" narcissism, which combines shyness and defensiveness, while "grandiose" narcissism is more exploitive and unempathetic. Yet both are attention-seeking. Here's how those two played out in one survey of owners of exotic pets: Vulnerable narcissists were more attached to pets, but only if they were exotic. But grandiose narcissists were more attached to traditional pets.

These results weren't dramatic by any means, but the authors' suggestion in the preamble to their report that narcissists desire "low maintenance relationships that enhance self-image" might be in play here. They continue: "[It] might lend itself to a desire for some types of exotic pets, *such as reptiles and insects*, as there can be limited interactions with such animals that, nonetheless, often evoke strong reactions from onlookers" (italics mine).[5]

That statement—wanting to startle others without having to work very hard for it—makes sense to me, but there are many such experiments, and the results always seem to me to be a little more uncertain than you'd like.

Sometimes, those pets defined as exotic seem anything but. I have owned "exotic" pets too: two red-eared slider turtles when I was a kid, and two lizards (of unknown species) when I was older. I'm pretty confident my motivation for the turtles wasn't to polish my image—in fact, I'm pretty sure it might have sullied it (almost as much as revealing that I had a microscope to look at the single-celled animals in pond water). The lizards? I don't think so, mostly because I don't remember showing them to many people, and I sure didn't talk about them at parties.

Jennifer Vonk at Oakland University has investigated narcissists and their exotic pets, and in one memorable instance, she went on to connect the dots between some unlikely topics: the Netflix documentary *Tiger King: Murder, Mayhem, and Madness*, the personality profiles of students who had watched it, and their attitudes toward exotic pets and unaccredited zoos—like Joe Exotic's. It's worth knowing that the documentary had ridiculously high audience approvals and 34.3 million viewers in the first ten days. Most scientists familiar with exotic species, especially tigers, found the show, or at least Joe Exotic himself, an appalling—not appealing—character, but if thirty-four million people are watching, appalling or not, there is a potential influence. That's what Jennifer Vonk was trying to find out.[6]

It was a difficult study because, ideally, she and her colleagues would like to have surveyed their study group's opinions of roadside zoos before *and* after they'd seen the series, but that was impossible given the suddenness of its enormous popularity. Who hadn't watched it? If they had been able to find people who had decided not to watch, that would likely have introduced an unwelcome bias.

It turned out that viewers of the series were generally unsupportive of exotic pets and roadside zoos, but not all of them. Those who had what the authors called "dark traits," including narcissism, low empathy, antagonism, and other antisocial attributes, ignored the many negative aspects of Tiger King and remained enthused about both exotic pets and roadside zoos. Dr. Vonk suspects that such people were more impressed by the human benefits—the fame and excitement—than the welfare of the pets.

Considering the negative impacts around the possession of exotic animals—capturing sometimes rare species in the wild, careless handling and shipping to consumer markets, ignorance of proper care, and accidental or purposeful release into novel environments—the picture painted of the kinds of people who want exotic pets, and the reasons they want them, tends to be a little depressing, but also inconsistent. However much we might believe that someone who owns a poisonous Gila monster is trying to tell us something about themselves, it might be hard to confirm that by answers on a survey. And it's also true that this is a bit of a one-sided picture—there are some defiant defenses of exotic pet ownership online, one in particular by Melissa A. Smith.[7] But there is some objective evidence too.

A survey involving exotic pet owners from thirty-three countries revealed that there was a general concern for the welfare of exotic animals destined for pethood. There was a preference for captive-bred versus wild-caught and for animals that are common in the wild. In addition to these attitudes supportive of conservation, people in the study rated "attachment, affection, nurture, as well as curiosity and being passionate about the species" as important. On the other hand, the respondents rated unusual colors or appearance highly. Any attitude that encourages rarity runs the risk of selectively removing certain kinds of individuals from the wild and breeding them intensively for the market.[8]

Another small study from Russia classified exotic pet owners into four groups: lifesavers, accidental owners, those seeking new experience, and collectors. Looking at those categories, you don't see a preponderance of people who are trying to enhance their personal image; choice seemed instead to be motivated by empathy, either for an animal or a human. Lifesavers stepped into situations where there seemed to be no future for an animal, whether a fox from a fox farm or an abandoned gray wolf. Accidental owners usually were taken by surprise: One woman acquired a tortoise from friends who fled the Russia-Ukraine conflict. They had owned the tortoise for sixteen years. New-experience people were a broad range of types, from formerly rural residents who had grown up with animals and now sought them again to a woman who claimed she would never again have birds because of their "biological needs." Instead, she had Egyptian fruit bats (!). The final group, collectors, were a mixed bag. Most of them had an attitude not unlike stamp and coin collectors, where their interest wasn't so much in the personality of the animal or the bond with it as in its unique features, its one-of-a-kind quality. Most collectors said they preferred captive-bred to wild-caught animals, although only one said that for ethical reasons, all the others because such animals were less aggressive and usually disease-free. True collectors.

This group, although it was only twenty-seven people in total, gives a much more textured view of people who own exotic animals. Maybe it's because it's Russia, not North America, maybe because the circumstances of acquiring the animals is so diverse, but together they defy easy labels. And when you look at the personality studies I mentioned earlier, the same is true. Yes, there are tendencies, especially with people who have elevated scores on narcissism, to gravitate to exotic pets, but overall, there's not a huge amount of commonality.

– PART V –

Downsides

– CHAPTER 18 –

The Outdoor Cat

If you were to trim the "outdoor cat" debate to essentials, it would be that cats shouldn't be outside, because they kill wildlife, versus cats must be allowed outside because it's their nature. Many of you have an opinion on this even if you don't own a cat, but this stripped-down version of the discussion ignores some of the most crucial issues involved.

Just about every cat owner who lets the cat out knows that, from time to time, their pet will bring home a sampling of their hunting prowess—a mouse, a bird, a lizard, whatever. But that in itself doesn't tell you much; it certainly doesn't make the hunting habit an "apocalyptic" threat to birdlife or worthy of a "moral panic," both terms you can actually read in articles about this.

That views are polarized here is not new. The distinguished ornithologist Edward Forbush wrote this in 1916: "The pet of the children, the admired habitue of the drawing-room or the salon by day, may become at night a wild animal, pursuing, striking down and torturing its prey, frequently making night hideous with its cries, sneaking into dark, filthy, noisome retreats, or taking to the woods and fields, where it perpetrates untold mischief."[1]

Nobody today puts it in quite those terms, but you get the feeling that the puzzle of what to do about cats and birds excites similarly deep feelings.

Downsides

The cat is the major player, but there are different kinds of cat, and each brings a different issue to the table.

So, first, the 100 percent *in*door cat. I've had cats like this that live their entire lives as if outdoors were an inaccessible exoplanet. On the positive side, such cats are not exposed to the myriad dangers of the outdoors, like predation (coyotes, eagles), infection by parasites, attacks from other cats and dogs (and humans), and death by car. The risks of contracting diseases, like rabies, could be on that list too, although they're not as significant.

Such cats can expect a much longer life. The most-quoted estimate is ten to twenty years for indoor cats, two to five for outdoor. But is that extended life a happy, fulfilled life? This is where the arguments drift into territory less well described by data. For instance, in an article called "A Defense of Free-Roaming Cats from a Hedonist Account of Feline Well-Being," Cheryl Abbate, a philosopher, argues that the indoor environment cannot provide the stimulation and experience necessary for a cat.[2] Abbate argues further that the owner has the responsibility to provide that opportunity—that failing to do so would amount to a moral failing. In her words, "especially pleasurable pleasures and rewarding experiences are available to felines only when they roam outdoors." Territorial behavior is a crucial feature of being a cat. In her view, cat owners (she calls them "guardians") have a moral duty to allow their cats access to the outdoors, as this is a crucial part of ensuring that the cat has the opportunity to "flourish." The philosophical term, attributable to Aristotle, is *telos*, which you could say is an animal's essence. Another philosopher, Bernard Rollin, borrowed a lyric from Jerome Kern and Oscar Hammerstein II to put it more memorably: "Fish gotta swim, birds gotta fly." He also wrote, "Violation of *telos* may be more significant to an animal than physical pain."[3] I do get hung up on the "may be." How do we know?

Abbate points out that she is not advocating that every cat must be allowed to go outside, wherever or whenever, acknowledging that

The Outdoor Cat

high-traffic areas and areas with large predators at nighttime are not safe for cats. However, in the absence of those cautionary circumstances, Abbate asserts that owners have a duty to let their cats roam.

If you agree, then you are accusing tens of thousands of unknowing indoor cat owners of moral failure. Knowing that, they might respond with questions like: What if I provide an outdoor-like environment indoors, with food puzzles (objects filled with food with small holes providing access) located differently every day, access to elevated places (including a perch with an outdoor view and room for only one cat), scratching posts in a favored area, and a selection of toys in constant rotation so the cat doesn't lose interest? Oh, and a clean, large litter box? Definitely sounds like more trouble than just letting the cat out!

Abbate has an answer for this: "While I grant that felines in sufficiently enriched indoor environments have lives worth living, I deny that they have *well-being*." Well-being in this case being something better than "just getting along," something more like "pleasure." (Exactly how to tell if a cat is truly experiencing pleasure is outside this analysis.) However, if the outdoors is indeed the ultimate place for a cat to be a cat, if it's immoral not to allow for that, then how? That introduces Cat #2.

Cat #2 (sometimes called an "inside/outside cat") goes out—sometimes. Many evenings, sometimes a morning or an afternoon here or there. What do we want to know about Cat #2? There's a list: What does the average outdoor cat do when outside? Is there even such a thing as an average outdoor cat? Can we identify which are the most prolific hunters? What prey are taken and in what numbers? And finally, after it's all factored in, how significant is cat predation on wildlife, especially birds?

Are cats skilled hunters? It depends. In one of the all-time great studies, conducted by William G. George, the subjects were his three cats, a mother and her son and daughter. They lived in the house, were fed in the house, but were free to range over 6.9 hectares (17 acres) of

Downsides

fields and 1.2 hectares (3 acres) of woods. As far as George could tell, they brought back every single kill they made, depositing it either on the lawn or in the house. Despite the fact that high-quality cat food was available to them every day, they hunted relentlessly for twelve hours or more every day, all year, except for winter, when hunting activities slowed dramatically. Five of every six winter days, the trio of cats caught nothing. But in a year (each of four years), they still hauled in an average of 483.5 vertebrates (a large variety of mammals, but also birds, reptiles, and frogs) and 286.4 mammalian fetuses belonging to slain pregnant females. Obviously, well-fed cats still hunt, and these particular cats were hunting machines.[4]

But these prodigious hunters may have had unusual advantages, like twenty acres of land available for hunting and plague-level populations of small mammals. By contrast, two studies that used either radio collars or "kitty cams" did find lower kill rates but, importantly, also showed that in cities, estimating kills from the bodies that turn up at the doorstep is not reliable. When cats' movements outdoors were recorded for analysis, it turned out that often they consumed the prey on the spot or abandoned it.

Based on corpses that were brought home (and kept in the freezer by the humans), the local cats recorded fewer than two kills a month, but that number more than doubled if the data from the trackers was included. In this study, prey included more small mammals than birds, and the cats, despite living near a forest preserve, spent much of their time in yards or just at the forest edge—unlike George's. Significantly, in these studies at least, cats appeared to have no discernible impact on the numbers of rodents.[5]

Two studies leave way too much room for speculation. Another study tracked 925 cats from six countries, and researchers found that while the cats' home ranges were small—only three of them covered more ground than a square kilometer (a little more than one-third of a square mile)—the monthly average of 3.5 prey items reported by the cats' owners ranked them about equivalent to wild predators, but significantly more destructive when their much more highly concentrated populations were taken into account. And these numbers were based on the notorious underestimates provided by prey brought home.[6]

And finally, to add some color to this picture, Norwegian researchers studying the movements of ninety-two locally owned cats were moved to describe what they saw as "a dense predatory blanket that outdoor cats drape over and beyond the urban landscape." They call this blanket the "catscape," a landform that prey animals have to navigate.

Some studies have shown the shape of the catscape matters. In one, where cats had access to nearby forest, most kills were made at, or just inside, the forest. The authors suggested that maintaining a cat-free zone in the vicinity of the woods, a DMZ, could dramatically reduce native mammal and bird fatalities.[7]

You can interpret the data from studies like this either as diminishing the purported threat to wildlife by cats or dramatically reinforcing it. How could it possibly cut both ways? You could point out that, yes, there's killing, but there is not a catastrophic number of prey. On the other hand, that's exactly the issue. These studies are limited in scope. What if you scale those numbers up?

The sticking point in this debate isn't so much the detailed numbers of prey caught—or likely caught—by individual cats, but how to extrapolate from those numbers to a general impact on wildlife. Each small group of cats that's been studied is unique. Depending on which study you choose, extrapolating from one or the other could result in a dramatically over- or underestimated threat. And as the authors of the Norwegian study pointed out, before their catscape project, "We are not aware of any study that has attempted to track *all or the majority of cats* within a neighborhood with a typical cat density. This is a surprising gap in information, as the local ecological impact of pet cats is attributable to their sheer numbers" (italics mine).

This is a serious deficit in the data. Most studies that attempt to calculate any sort of continental, let alone global, total on the numbers of birds and animals killed by cats scale up the figures provided by local studies. The more such studies, the more reliable the result, but the differences among those studies are substantial enough to weaken that data.

But that doesn't make them meaningless. There are such large-scale studies, and they're heavily quoted. In recent years, the most dramatic was a report published in 2013 in the journal *Nature Communications*.[8] A team of scientists from the Smithsonian Conservation Biology Institute and the US Fish and Wildlife Service claimed that in the United States "free-ranging domestic cats kill 1.3–4.0 billion birds and 6.3–22.3 billion mammals annually. Un-owned cats, as opposed to owned pets, cause the majority of this mortality." The authors point out that these numbers are dramatically higher than previous estimates and argued that kills by cats are greater than any other source of wildlife mortality. For birds, cats are worse than window strikes and collisions with communication towers. For birds *and* mammals, cats outkill vehicles and pesticides.

This publication poured gas on the embers of this debate. In fact, this is when the words "apocalyptic" and "moral panic" first appear. And those overheated terms were only the tip of the iceberg. Here's one:

> If even half the U.S. Fish and Wildlife's energy and funding was geared toward spay/neuter and educational programs, this problem wouldn't be an issue. But that, of course, would result in a lot of well-connected, taxpayer-reliant academics being forced out of work.[9]

There were some critiques worth noting, especially those who reminded everyone that the authors admitted they really had no handle on the numbers of unowned—that is, feral—cats, but they were sure they accounted for the majority of the kills. Feral cats might compose nearly three-quarters of all hunting cats, but there are no hard numbers. They are, by definition, nearly impossible to track.

But even with this gap in knowledge on their side, critics of the report had to admit that cats do kill birds. Their concern seemed to be that public acceptance of these huge numbers would turn people against cats. Maybe even advocate culling feral cats. That was *not* a proposal put forward by the scientists.

Ten years after the original *Nature Communications* publication stirred things up, the journal turned up the heat in 2023 with "A Global Synthesis and Assessment of Free-Ranging Domestic Cat Diet," an international study that generated these quotes: "Collectively, our findings demonstrate that cats are indiscriminate predators and eat essentially any type of animal that they can capture at some life stage or can scavenge" and "Our database is likely to grow markedly in the future and represents only a fraction of the true magnitude of species consumed by cats globally."[10] The authors identified 2,084 species killed by cats, of which about a sixth were of "conservation concern."*

Temperatures run high over studies like this. Alley Cat Allies hit back by highlighting a weakness in the study: that there was no distinction between prey that had been killed and prey that had been scavenged. Labeling all such corpses as victims of predation is an exaggeration. But then, apparently having run out of critique, the authors, perhaps fearing the article would lend new support to culling feral cats, spent some time attacking news reports of the study, including UK tabloids, likely one of the easiest targets ever.[11]

---

* In referring to the 2023 *Nature Communications* paper, Alley Cat Allies had this to say: "The article refers to a 'strong impetus to advance policy and management initiatives that seek to reduce the impacts of free-ranging cats,' but what, exactly, are those policy and management initiatives? Alley Cat Allies believe they imply 'trap and kill' programs."

Downsides

Then there's the feral cat, the most important but also the most mysterious, its gigantic numbers worldwide of uncertain impact because of the relatively scant knowledge we have of their hunting habits. Those numbers, rough estimates only, are about five hundred million, of which close to one hundred million are thought to live in North America. This is a cat that is not fed, housed, and rarely even seen by humans, let alone petted. Obviously, unlike the house cat, which is fed regularly and still hunts, feral cats need to hunt to survive, and that need, combined with their numbers, means that the risk posed to wildlife by household cats, whatever it is, is dwarfed by the impact of feral cats.

Household cats or feral, the barrage of statistics is everywhere. And while the United States leads comfortably with 2.4 billion annually, Canada, Australia, and the United Kingdom report numbers from the tens to hundreds of millions, suggesting this is a global issue—although, remember that these numbers are derived from much smaller studies, and scaling up inevitability brings with it uncertainty.

Where there is certainty is in the deplorable record of carnage that has resulted when cats are introduced to islands—by humans, who have also facilitated the introduction of other agents of extinction like domestic dogs and rats. As reported by a publication in 2016, cats have been involved in sixty-three extinctions: forty bird species, twenty-one mammals, and two reptiles. Only rats (three different species) have been involved in the extinction of more, seventy-five in all. And the reported totals are sure to rise, as there were twenty-three critically endangered species, possibly extinct, not included in the data.[12]

The island story is almost always the sad combination of a skilled predator and a naive prey, with scant protection on the landscape. However, its relevance for the indoor/outdoor cat controversy is minimal except to cast a lurid light on cats as hunters, enabled, of course, by humans.

This last point is sometimes understated. Humans are at the root of many of these catastrophic extinctions. For example, Marion Island in the Indian Ocean: First, humans inadvertently introduced mice to the island, which started killing off the birds, unused to predators as they were. Then, in 1948, five cats were introduced to kill the mice. They went feral and by the 1970s were killing nearly half a million birds every year. Cats were "eradicated" in 1991 and now the mice are back. Now the plan is to cover the entire island with a rodenticide.[13]

Cats kill birds, mammals, and reptiles. When completely free to do so, they can wipe out a susceptible population. But household cats that spend only part of their time outside are certainly not the same as feral cats on an island, or even those at the edge of a forest. Nonetheless, they do kill, and the issue then becomes how many. Is it an overwhelming number of birds and mammals? If not, is it still enough to do more to protect wildlife?

What would that be? Keeping cats indoors is a 100 percent effective solution to the problem that many—as we've already seen—would reject. Creating outdoor cat spaces—"catios"—at least provides some elements of the outdoor experience, but whether that's fulfilling enough, I'd leave to the philosophers. Outdoor fencing that limits a cat's activities to the owner's yard might be the best solution, especially given that occasionally an unwitting bird might actually provide hunting experience. The effectiveness of having outdoor cats roam free wearing bells, bright colors, or other devices is dependent on consistency, and that unfortunately rests with the human. So it's far from perfect.

Those are techniques for household cats with active owners. Feral cats are a different story. This is where the pro- and anti-cat factions collide over a single remedy: TNR, or trap-neuter-return. It sounds straightforward—trap feral cats, neuter them, and let them go, confident they will not reproduce. Over time, they all die, and that particular colony of feral cats is gone.

Unfortunately, TNR doesn't work. Feral cats have a number of attributes that make them bad customers for this. First, it's very difficult to locate them all. Almost by definition, they're elusive. Leaving even one or two fertile cats behind is a guarantee that the colony will survive. Second, even if you could locate the ones that are in the colony right now, that doesn't account for the other cats that are likely to infiltrate from somewhere else. The population of a feral cat colony is fluid, with constant immigration and emigration. It's estimated that 75 percent of any particular colony must be neutered for TNR to work. Examples of such success are rare.

TNR brings out the extremes. One veterinarian, David Jessup, in a report called "The Welfare of Feral Cats and Wildlife," recorded his personal experiences of TNR: "I have personally seen multiple feral cat colonies . . . where various levels of TNR (from casual to serious efforts) have gone on for many years. None of these efforts, by themselves, eliminated the feral cat population." Jessup went on to articulate the reasons for TNR's failures: The initial number of cats must be relatively small (thirty to forty); there can be no immigration; all or nearly all females must be captured and spayed; and efforts must be "early, intense and prolonged." Such conditions don't last long enough.[14] Computer simulations reveal that a population of two hundred feral cats will never be eliminated by TNR. It will decline, then level off, but never decline to zero.

Perhaps a small part of opposition to TNR is based on the neutered cat never being able to achieve its psychobiological need for mating and raising offspring. Support for TNR comes largely from groups like Best Friends or the Humane Society of the United States, whose motivation is the avoidance of any euthanasia—period. The same view motivates these organizations' eagerness to discredit studies that estimate kill rates for entire countries (those studies that admittedly have uncertainties).

Some of the arguments against TNR feel like a reach too far. Like the claim that songbirds, especially in Europe, have evolved in the presence of cats for centuries and so have learned to avoid them. That's conceivable in Europe, but North America? Dense populations of feral cats haven't existed long enough there. The same holds for the argument that, for instance, the European blackbird is a more recent arrival to urban life than the cat, so the cat shouldn't be viewed as an invasive species. But humans didn't bring blackbirds to Europe. Nor do they supplement the blackbird population by abandoning their own pet blackbirds. Human involvement must always be separated from natural migrations.

So let's leave it squarely in the humans' domain. In a paper published in 2020, the authors examine the laws around this issue and conclude: "Many national authorities around the world are currently required, *under international law*, to adopt and implement policies aimed at preventing, reducing or eliminating the biodiversity effects of free-ranging cats in particular by a) removing feral and other unowned cats from the landscape to the greatest extent possible and b) restricting the outdoor access of owned cats" (italics mine).[15]

Somehow, I don't see this resolving the issue.

– CHAPTER 19 –

Beyond Cats

Step outside the room where the outdoor/indoor cat opponents are going at it, uh, tooth and nail, and it's strangely quiet. The insults, inflammatory language, and bitterness are largely absent from the issues of other negative impacts of pets, starting with dogs but extending to exotics.

Dogs are the world's most abundant carnivore, and of the world population of nine hundred million, 75 to 80 percent are thought to be semi-wild. The numbers are therefore similar to the numbers of feral cats.

The discussion isn't as heated at least partly because dogs are commonly leashed, especially in Europe, much of North America, and sporadically in Asia, while cats on leashes are still a rare sight. Yet the worldwide numbers reveal millions of unhoused dogs that must occasionally come into contact with wildlife.

We're not debating whether dogs are capable of carnage. A single German shepherd in the Waitangi Forest Reserve in New Zealand was probably responsible for the death of something like five hundred kiwis out of a total population of nine hundred.[1] On some of the islands in the Galápagos, marine iguanas are under constant predatory pressure from dogs, whose numbers on those islands are growing dramatically. In case you thought the Galápagos were pristine nature reserves, there have been dogs on the islands for more than a century, and one survey showed that the number of dogs on Santa Cruz Island alone increased by 55 percent from 2014 to 2018.[2] To some extent, these examples are unusual, combinations of particularly vulnerable prey (iguanas have developed no defensive behaviors with respect to dogs, nor have kiwis) and easy access. But where such combinations exist, dogs will take prey, or mess around with wildlife in other ways.

One recent estimate is that dogs have contributed to eleven extinctions and threaten 188 species around the world. The extinctions include the thick-billed ground dove, the Tonga ground skink, and the New Zealand quail. The 188 threatened include ninety-six species of mammals, seventy-eight birds, twenty-two reptiles, and three amphibians. (The authors claim these numbers are likely underestimates.)

The canine threat is 80 percent predatory, with disturbance of wildlife, disease transmission, competition with other predators, and even hybridization with wild relatives (thereby diluting their genomic uniqueness) making up the rest. The high-risk areas for most of these threatened species are Mexico, South America, and Southeast Asia. This might account for the relative silence of the loudest voices in the indoor/outdoor cat controversy, who are concentrated especially in the United States.[3]

While it would seem that islands, forest reserves, and the Galápagos are unexpected places to find dogs, if there are humans, there are dogs. And wherever there are dogs, they will, if given a chance, threaten wildlife. And we're not just talking remote areas of New Zealand or tiny, isolated islands in the Pacific. It's true even in and around heavily populated areas. Beaches are a good example.

Dogs running free on beaches threaten shorebirds. Dogs displace them, or simply kill them and their nestlings. This is not a secret. Why does it continue to happen? As always, humans. Communicating that risk to the public and getting them to do something about it—like keeping dogs on a leash—have pretty much failed.

Some of the best data has come from New Zealand and Australia.

Downsides

In both places, there are wildlife refuges that combine conservation of shorebirds (some of them the rarest in the world), with sports fishing, surfing, and dog walking. Surveys done in these places reveal that when it comes right down to it, dog walkers feel it's their right to be able to take their dog to the beach. Some of those also believe that for a dog to truly enjoy itself, he/she has to be free to run (shades of a cat's right to roam around outside). Many would go so far as to assert that their dog has as much right to be free on the beach as the shorebirds do.

In one study in New Zealand, nearly half of dog owners who let their dogs run on beaches knew the dogs posed a threat to birds. Even those who thought dogs should not be allowed on the beach weren't concerned for nesting birds, but for their own personal experience, which they felt was diminished by the presence of dogs. Dog owners who wanted to be able to take their dog to the beach despite the presence of vulnerable birds argued that "my dog wouldn't hurt a flea." In (partial) defense of some of those humans, they might not realize that even the sight of a dog running along the beach, paying no particular attention, is enough to incite fear and evasive action in some birds. Even a dog on a leash can spook a bird away from nest and nestlings.[4]

Of course there are always suspicions that human compliance with regulations is directly related to the likelihood of getting caught.

One of the reports summarized the sad situation: "With seven years of intense education, awareness campaigns, and enforcement of leashing laws in one National Park, compliance peaked with only 22 percent of dogs leashed."[5]

And it's not just shorebirds. Although the disruption of their nesting has been the focus of much of this research, there's also evidence that the mere presence of dogs on a beach during the day inhibits nighttime scavengers, mostly mammals, from cleaning up the carrion that inevitably builds up—a subtle but environmentally crucial operation.

No matter where you stand within these controversies, cats and dogs confront wildlife in various ways. But the influence of their presence doesn't stop there. Our favorite pets have other, far-reaching negative environmental impacts.

Here's a delightful stat: All the feces produced by dogs and cats in the United States for a year is roughly equal to the total amount of garbage produced by the state of Massachusetts, population about seven million. (These calculations were based on 2015 numbers, and while the Massachusetts population has grown by a few hundred thousand since then, it's a safe bet that the number of dogs and cats in the USA has grown more.)[6]

I've backed into this story, which is really about food consumption, not its downstream outcomes. Greg Okin of UCLA did a painstaking rundown of the basic elements of pet life that have environmental consequences. Feces are one category: That Massachusetts comparison is a more dramatic way of saying that fecal production by dogs and cats in America is *30 percent* of the human equivalent. Put that way, I understand why our dog walker always used to let me know if Robbie had "done his business." Every dog's contribution is essential!

How can you figure out what has to be consumed, energy-wise, to generate those landfills of dung? This is something we should know. Okin worked through a long analysis starting with the essentials of pet food (protein, carbohydrates, fat, water, and ash), allowing for differences between wet and dry food (dry food outsells wet food for both dogs and cats) and differences between premium and gourmet versions (there's usually more meat in "gourmet"), and found that 33 percent of cat and dog diets comes from animals (chicken, beef, pork, fish, etc.). Significantly more than the human 19 percent.

When all the numbers are combined, of the total energy consumption by pets *and* humans in the USA, dogs and cats are responsible for 20 percent. Doesn't sound like much, but 140 million people could live on an energy-equivalent diet. Year-round, forever and ever.

The fact that pets consume more meat in their diet than we do

Downsides

usually prompts the argument "Yeah, but that's meat no one would eat: lips and tongues and stuff. If it didn't become pet food, it'd become trash." Of course, if that meat was going to be thrown out, not eaten, it shouldn't be counted against pets as *extra* consumption, but Okin counters there are plenty of ingredients in pet food that could be processed for human consumption. That undermines the argument that pets are eating only byproducts. But not only are there already pet foods containing products that we eat, more are on the way as consumers drive the popularity of premium and gourmet versions of cat and dog food.

The central issue is the pressure from pet food, especially beef, on the environment. The beef industry is being watched very carefully to see how fast its troublesome greenhouse gas emissions can be reduced, at a time when increasing wealth worldwide has more people demanding more meat. Meanwhile, pets are on that same upward climb.*

None of the things you'd suspect of being pressure points seem to offer much. Change pet owners' buying habits and encourage cheaper, less meaty foods? That's reversing an existing trend. Substitute new protein sources, like insects (a win-win-win if you ask me)? Feed your pet less? Getting anyone to listen?

Greg Okin says that the food production system needs to be dramatically overhauled to shrink the consumption of meat products. He also had this suggestion: "Reducing the rate of dog and cat ownership, perhaps in favor of other pets that offer similar health and emotional benefits, would considerably reduce these impacts." They would, but how is that going to happen? Okin suggested alternative

---

* A ray of hope here is that in July 2024, a UK company, Meatly, was approved to produce its lab-grown chicken cells for pet food, the first time anywhere such approval has been given. Lab-grown meat eliminates many of the environmental concerns around meat, but is still in the early stages. It is too expensive to be practical at this point, and beef industry lobbyists are actively opposing it. So we will see.

Beyond Cats

pets, all herbivores: birds and hamsters for the household, horses for outdoors.* But making any sort of dent in the cat- and dog-owning world—at least in North America—you'd have to wield something more powerful than "they are eating too well." It's hard to imagine the era of the dog and cat winding down, but who knows what sort of pet environment will exist in the year 2100? Maybe the greenhouse gas connection to beef production will be over—one way or the other.

Dogs and cats, by sheer numbers, exert pressure on the environment. But they're not the only players. So-called exotic pets are shooting up in popularity globally, their spread outracing the knowledge needed to take care of them. They're all sizes, many species: axolotls, tegus, canaries, fire ants, ball pythons, zebra finches, leopard geckos, tortoises, chameleons, and thousands more. Some of them are dangerous even to their owners, but the biggest risk is escape or abandonment leading to the establishment of an invasive species. The classic story is the Burmese python in the Florida Everglades. Abandonment or escape, no one really knows how it started. The first Everglades python was seen in 1979, and it took a few decades for the population to grow to a point where its impact on native wildlife could be measured.[7]

A study published in 2012 showed by counting mammal roadkills at night on roads near the Everglades that between 2003 and 2011, raccoons there declined by 99.3 percent, opossums by 98.9 percent, and bobcats 87.5 percent. That they were killed by pythons is the only reasonable explanation—the farther from the pythons' range, the more abundant these animals are. And that's just three mammals: Pythons eat birds, attack alligators (sometimes swallowing them), breed with speed, and really have nothing that threatens them except the South Florida python removal agents.

The size of the snakes—five meters (fifteen feet) or more—and their

---

* I'm interpreting Okin's suggestion of pet birds to mean fruit, vegetable, and seed eaters. Yes, many birds are carnivores; most pet birds are not.

fecundity—up to one hundred eggs at a time—are impressive, as is their predation of a huge swatch of wildlife. But Burmese pythons, like most invasive species, have subtler effects too. For instance, in the Everglades, the cotton rat is flourishing. Pythons concentrate on larger prey, including those that would have preyed on the rat. Ecosystem disruption comes in many guises. And predicting how this will all turn out is impossible. By now, pythons are settled in 2,500 square kilometers (1,000 square miles) of South Florida, including all of the Everglades.

Burmese pythons attract the most public attention, but exotic pets around the world are a never-ending supply of invasive species. To be fair, most lists of invasive species are dominated by plants and insects, with the insects largely introduced accidentally as stowaways on trade goods. But the exotic pet trade looms as a source of vertebrate species that might outcompete or prey on native species. Although there is a long history of introducing species deliberately for all the wrong reasons (a colonial attitude bestowed the starling on North America, and the bad idea of importing a foreign species to deal with a local problem brought the English sparrow to North America and the cane toad to Australia), the exotic pet trade has suddenly asserted itself, especially in South America and Asia, as the major force behind the movements of animals around the world, legal or not. With millions of animals on the move, there will be a steady stream of escapes, willed by humans or not. Invasions of black-and-white tegus, lionfish, red-eared sliders, Nile monitors, and green iguanas are the result (there are many more).

How often does this happen? Florida is an outlier: Its hospitality to invasive exotics is stunning.* One study revealed that of 140 non-native reptiles and amphibians now living in Florida, a stunning 85 percent arrived via the pet trade. This is off the scale compared with a recent international study showing that a little more than 50 percent

---

* Number of native lizards in Florida: 16. Number of introduced lizards: 43.

of invasive vertebrate species have been provided by the pet trade. That seems like a reasonable figure, making Florida an anomaly.[8]

The Florida reptile/amphibian data absolutely destroys the old "tens" adage that 10 percent of animals arriving in a new geographic area find themselves freed, 10 percent establish themselves as self-sustaining, and 10 percent of those become pests. Not all of Florida's 85 percent have become pests yet, but Florida is now number one in the world for having non-native reptiles and amphibians.

Escape most often results from human carelessness or ignorance of a species' skill and ingenuity, but release is deliberate. To reduce the incidence of both, we need a better understanding of the exotic pet species and their potential impact on the environment. Release or abandonment is a willful act on the owner's part, unlike escape, and it is usually triggered either by the pet growing to an unmanageable size or getting too old. That's one of the things about exotics: People buying dogs and cats know roughly how many years they'll be alive and what size they'll be. But exotics? Are you really ready to own one for twenty-five years? Will it soon need its own room?

The Florida statistics are attention-getting but not typical of the rest of the country, or the world. However, invasive species are the second-biggest force behind loss of biodiversity after habitat loss, and the damage costs the US more than $100 billion annually. The author of one analysis went nonscientific for a moment when evaluating the scale of the exotic pet trade: "The market in exotic pets has grown considerably since the 1970s, and the volume of vertebrate animals that are traded worldwide is shocking, even to relatively seasoned invasion biologists."[9]

It's difficult, if not impossible, to gauge how much of the international pet trade is illegal, but one estimate a decade or so ago suggested it was more than 50 percent of a multibillion-dollar industry. The illegal part is opaque to analysis, but some characteristic features of the pet trade have been unearthed, including the fairly unsurprising one that the more individuals of a species are imported, the more likely that species is to become invasive, simply a matter of sheer opportunity.[10]

More interesting is the notion that the characteristics of a species that make it appealing to pet owners are the same characteristics that enable it to become invasive. I should say invasiveness does not necessarily correlate with appeal across the board, as Pablo Escobar's hippos would attest—though they weren't pets. It is tricky to make the connection between pet appeal and the ability to become invasive, because many invasive vertebrates have been traded for decades, long enough that some species, which started out with a small chance of invasive success, nevertheless established themselves in their new surroundings. In that case, time, not adaptability, could be the crucial factor. To get a handle on this, scientists from the University of Lausanne analyzed the trade in pet ants. (Pet ants? See chapter 12.)

Why ants? Unlike vertebrates, they haven't been a major factor in the exotic pet trade for long enough to have become invasive, a process that usually takes decades. So if the ants being traded most heavily now—520 species of ant were sold online between 2017 and 2022—are also those with well-known attributes for being invasive, that's a threat. And it seems to be true: Ants with invasive potential are six times more likely to be part of the pet trade than noninvasive species. Two significant features favoring invasiveness are a large native range and a generalist role. Both make sense as influencers of the pet trade: Bigger ranges increase the likelihood that a species is captured for trade, and generalists are able to survive better in the variable conditions provided by pet owners.

This finding doesn't exactly complicate the issues of the exotic pet trade so much as underline them. Unless some sort of controls are executed, the situation is likely to build on its growing trend and get worse.[11]

After all this—dogs killing kiwis, Burmese pythons killing everything in reach, pets eating an amount of meat that would sustain millions of people—has the thought crossed your mind that maybe pets just aren't a good idea? Some have had that very thought.

There are a variety of opinion pieces online that appear to be advocating for the elimination of pet-keeping, only to stop short of that and merely argue that we need to stop the many abuses involved, like capturing animals from the wild to bring into the pet trade, running puppy mills, and careless ownership. Underlying the objection to these practices (that are already seen by most as bad) runs a philosophical thread that we, by the very ownership of other animals, are automatically denying them the best life. Again, as I wondered about the argument that cats must be able to be outside to be fulfilled, how do we know what or how much we're denying the family dog or cat?

Some ethicists, however, come straight to the point: "We oppose domestication and pet ownership because these violate the fundamental rights of animals." Those words were written by Gary Francione and Anna Charlton of Rutgers University.[12] They argue that by "owning" animals—regardless of whether we call it that or something like "companionship"—and dictating their lives, we fail to acknowledge their rights.

Francione and Charlton live with six rescue dogs, but argue that they wish the dogs had never been born. "We love our dogs, but recognize that, if the world were more just and fair, there would be no pets at all, no fields full of sheep, and no barns full of pigs, cows and egg-laying hens. There would be no aquaria and no zoos." As unlikely as that looks—and it is *very* unlikely—I like it because it puts a stake in the ground, even though that stake might be far away from where we are right now.

An article by Troy Vettese in the UK's *Guardian* put it much more directly and aggressively.[13] He focuses not on the denial of pets' rights so much as the cruelty we inflict on them and the misery that results. In "Want to Truly Have Empathy for Animals? Stop Owning Pets," he cites many examples of pet abuse: exotics being captured and dying on their way to market, the miserable life in puppy mills, short-lived inbred animals, dogs being hit by cars, parrots defeathering themselves. And even this: "It is hard to fathom the boredom of pet fish." It's as hard as imagining life on the reef, the nonstop search for food and avoiding becoming food.

Vettese thinks we have reduced pet animals to "mere dolls" or "commodities," and made them that way by training (they are "broken" by that), inbreeding, and the damage to ecosystems caused by the production of food to sustain them. He has a solution: "We must collectively decide to shut down puppy mills, to spay and neuter pets and to support conservation programs that humanely capture feral animals."

That would do it! It's the "collectively" that stands in the way.

– CHAPTER 20 –

Eat . . . or Be Eaten

I'll probably eat you when you are gone
And you've probably known this all along
I'm a cat . . . and that's that
—Lyric by Trevor Day

Eating and/or being eaten can stir up the most powerful of emotional responses. Cannibalism is an obvious example, so powerful a topic that it has often been used to denigrate groups of people accused of the act, regardless of whether it was true or not. It is just as disturbing if we're talking about pets. This chapter addresses the issues attached to two different questions: "Would you eat your dog or cat?" and "Would your cat or dog eat you?" If either of these topics makes you feel something between seasick and carsick, just skip the next few pages.

I had two decisions to make before writing this: Should I tackle the subject at all? I decided yes because it has implications both for philosophy and forensic science—it is more than just sensational. Then I had to decide, what to cover first? I chose "eating your pet" because it is really an idea, not an actuality, at least in most cultures. "Your pet eating you" is grimly based on real situations.

The question "Would you eat your pet?" comes with some qualifiers. The most important is that it presupposes that you didn't kill your pet to eat it—it died accidentally, and you were then faced with the issue of what to do with it. And although the question could apply to any pet ("Would you eat your ball python?"), discussions around the issue are usually focused on dogs or cats, understandably so because the emotional issues are sharpened by their familiarity and the added likelihood that many of you who think about it might own or have owned one.

There are two popular versions of the question. One, offered by Jeremy Stangroom in a book called *Would You Eat Your Cat? Key Ethical Conundrums and What They Tell You About Yourself*, goes like this: A woman named Cleo Patrick owns a cat named Hector, with which she is very close. They go to the supermarket and watch reruns of *Melrose Place*, and she reads to him at bedtime. Sadly, Hector does not see well and one day mistakes a lawn mower for a mouse and dies instantly. Sad, but Cleo had decided long in advance that if something like this were to happen, she would eat him—he would become one with her. And so she does: "Cleo sat down on the evening of Hector's death and ate him on toast, washed down with a nice glass of chianti." (The chianti is a nice touch.) She never regretted her decision.[*]

I imagine a large majority of those reading this are repelled by the idea, but Stangroom asks you, before you condemn Cleo, to consider that Hector wasn't killed to be eaten, that no one was harmed—neither Hector nor Cleo nor anyone else—and that Cleo was happy she had done it. So what's wrong with it?

If you think it's wrong, you are likely elevating your feelings about it—disgust, maybe, or at least significant discomfort—above a rational evaluation of the incident, and even though no one was hurt, you still think that something like this, though an entirely private act, can be morally reprehensible. (If you're vegan, you might argue that eating any animal is morally reprehensible, but that belief is often based on

---

[*] There was a similar incident on *The Simpsons* with Homer and a lobster.

the shunning of chicken, beef, lamb, and fish either because of the terrible lives these animals lead, the manner in which they're killed, the impact that the eating of them has on the environment, or all three. Such arguments do not apply here.)

Years ago, long before he was well known for his advocating cutting back cell phone use by children, the social psychologist Jonathan Haidt coauthored an article called "Affect, Culture and Morality, or Is It Wrong to Eat Your Dog?" in the *Journal of Personality and Social Psychology*.[1] The study asked whether in two different countries, Brazil and the United States, actions that would be considered disgusting by all could nonetheless be treated as moral or immoral, depending on the culture. Could, for instance, an act that harmed no one be thought by some to be immoral and others moral? Three cities were involved, Philadelphia in the United States and Recife and Porto Alegre in Brazil. Interviews were conducted in both lower and higher socioeconomic classes in all three cities, with the expectation that higher-income people tend to be more individualistic (and permissive) than moralistic. Adults and children were involved.

One of the questions on the survey read like this: "A family's dog was killed by a car in front of their house. They had heard that dog meat was delicious, so they cut up the dog's body and cooked it and ate it for dinner."

The study invited comment on several other stories, but reactions to the dog story were consistent with the others: Philadelphians, by and large (especially high-income individuals), did not think eating the dog was immoral, while Brazilians, especially those from low-income areas, thought it was. Given that this example, like the one with Cleo and Hector, harmed no one, harm is not necessarily the sole factor in decisions about whether an act is moral or not. Not just that, but moral decision-making will differ depending on culture.

This study highlights an essential difference in the making of moral judgments; sometimes they are rational, depending on considering

Downsides

factors like harm, and to whom, and sometimes they are emotional. I lean toward reason, not emotion, but at the same time recognize that stories like these can often depend on the detail of the storytelling. For instance, would you like to hear exactly how Hector and the dog were butchered? Or how they were cooked? Cut up into tiny chunks that rendered them unrecognizable in a casserole, or carved like the Thanksgiving turkey? There are points along the storytelling arc where nausea might set in—whether that would influence judgments of morality is another question.

In defense of gut decisions, aside from reason or logic, Leon Kass, once the chair of the President's Council on Bioethics, said, "Shallow are the souls that have forgotten how to shudder."

And a final word on the portability of morality, the idea that a moral thing in one place might be immoral in another: Eating dogs and cats is, broadly speaking, verboten in most countries, except if it were to be a lifesaving necessity, and except in Southeast Asia. Several Southeast Asian countries are untroubled by eating dog meat, maintain public dog meat markets, and obviously have adopted a different moral stance toward dog meat than much of the rest of the world. You might, if you're a dog-lover, cringe at the thought of dog meat producers experimenting with different breeds to find both the tastiest meat and the fastest breeders, because puppy meat is considered the most desirable. But before criticizing them, you need to remember how chickens, pigs, sheep, and cattle are treated elsewhere.

I'll now give you a second chance to flip forward a couple of pages if you're squeamish. If it wasn't easy for you to consider consuming your favorite pet, what about the reverse? Sadly, you die at home, unnoticed by family or friends, but accompanied by your dog or cat (or both). What happens?

Of course there are no hard and fast predictions that can be made. There are stories of a faithful dog staying beside the owner's dead body until it starved to death, but that's not always the case. In a column in

Eat . . . or Be Eaten

Psychology Today, Dr. Stanley Coren at the University of British Columbia searched for information on how likely it is that dogs or cats would start to eat the dead owner's body and couldn't come up with much scientific evidence, but he did talk to several first responders. The information he gathered was inconsistent, but more than one person thought that cats wouldn't wait as long as dogs before they started to chew at the body; that some dogs would wait until the body started to putrefy, then give in, although a famous case in Germany put the lie to that. A man died by suicide in a shed in his backyard, and by the time help had arrived (forty-five minutes later), his German shepherd had chewed on his face and swallowed what he had eaten. A half-full dish of dog food was nearby.

"Postmortem scavenging" it's called. Police in Australia suspected a sixty-nine-year-old man might be dead. When they opened the door to his place, thirty cats emerged. Much of his body had been eaten, and in one final grotesque flourish, a cat was perched inside his empty chest cavity. This was an unusual case—most of the time there are only one or two cats, and they confine their scavenging to soft tissue, like the lips and skin.

Although statistics are hard to come by, Roger W. Byard, a forensic scientist at the Adelaide Medical School, University of Adelaide, is sure that "scavenging is more common with dogs than cats." But he also says, "I don't trust either of them."[2]

Both dogs and cats tend to start at the face, then work down to the torso. Once they begin, it seems they don't stop, but that's the question: Why begin? The German example shows that even with food nearby, a dog might begin to chew at the deceased owner's face. There are some theories, but they seem far from definitive.

One—and this comes from the scientists who investigated the German case—is that perhaps the dog tried to wake the dead man by licking and nuzzling his face, but as that didn't get results, the dog became more insistent, even biting. The scientists label that as "displacement," an initially sensible behavior becoming exaggerated. If time passed and the dog ran out of available food, then hunger might exaggerate the behavior already underway. Some have suggested that the odors of putrefaction, inconsistent with the odor of a living person (especially one who's familiar), might trigger scavenging. And while Stan Coren didn't find much about the *likelihood* this is going to happen, there are many, many reports in the forensic literature.

Investigating such disturbing cases of scavenging has important implications both for health and safety and criminal investigations. While an accidental or sudden death raises, however unlikely, the possibility of scavenging, people who hoard pets run a much greater risk. Animal hoarding disorder is an officially recognized mental health condition. Despite the name, I'm not aware of a case in which the animals were not pets. People with it accumulate large numbers of animals (even hundreds), yet neglect them even to the point of death while failing at the same time to maintain their own health.

In most cases of pet hoarding, there's a central conflict: The hoarders believe they are doing everything possible to provide for and protect their animals, while most often the animals are unwell and undernourished. There's a broad range of hoarders: Some truly love their animals but become overwhelmed by the caregiving necessary; some actually have no empathy for their animals and allow them to suffer.

If the hoarder dies, there's a definite risk of postmortem scavenging. Interventions to stop hoarding are difficult—hoarders are convinced they're providing the best for their animals—but a greater recognition and awareness of cases of hoarding might help prevent scavenging.

Criminal investigations in which an examination of the body is necessary to reveal the cause of death are hampered by any scavenging that has already happened. In one case in Chile, the first examination of the body of a woman whose face had been chewed by a dog attributed the death to natural causes, but further examination revealed that she had been struck in the face during a robbery.

Eat . . . or Be Eaten

The images associated with these cases are disturbing, especially given the discomfort of knowing that the human and the animal were in most cases attached to each other. But the more evidence that comes to light, the better the understanding of the motivation of the animal—and that might lead to strategies to prevent cases like these.

One thing is for sure: The people who have examined even the worst cases of postmortem scavenging are open-minded about it:

Stanley Coren, University of British Columbia: "Personally, if I am dead, I have no particular need for my body and if it sustains my pets until they are rescued, that is fine with me."[3]

Carolyn Rando, forensic anthropologist, University College London: "I think we have to come to the conclusion that our pets will eat us. . . . It's just a fact of life."[4]

Roger W. Byard, University of Adelaide: "If it kept my old golden retriever going after I died . . . I'd be quite happy for it to have a feed."[5]

– PART VI –

Oddities

– CHAPTER 21 –

Road Trip

Holly, Howie, Bobbie the Wonder Dog, and the famous trio of Luath, Bodger, and Tao compose an unlikely group: All are said to have traveled some incredible distance—on their own—to return home. What distinguishes them? Luath, Bodger, and Tao are famous but fictional; the others actually performed incredible journeys.

Luath, Bodger, and Tao are the heroes of Sheila Burnford's 1961 novel *The Incredible Journey*, subsequently made into a hugely successful Walt Disney film of the same name, then again in 1993 into *Homeward Bound: The Incredible Journey* and *again* in 1996 into *Homeward Bound II: Lost in San Francisco*. The three animals, Luath the golden retriever, Bodger the bull terrier, and Tao the Siamese cat, having been separated from their owner, make their way through 480 kilometers (300 miles) of Canadian wilderness to return to their home. This has obviously been a story of enduring popularity, earning the reputation along the way of being true, but it is not. At best, Burnford based the story on other accounts of animals making their way cross-country to reach home. *The Incredible Journey* is not a true-life story of an actual trio of animals, but it's not impossible that the feat could be accomplished.

The claim that "it's not impossible" requires some evidence, and there is plenty. For instance, there's this account, by W. D. Harry, of an unnamed collie in the September 22, 1922, issue of *Science* magazine: "One of our neighbors directly across the street moved by rail to Denver, taking this dog less than a year old with him on the train. In less than a week he was back at the old premises and barking joyously as ever."

Mr. Harry (writing from Canon City) prefaces that account of the dog's return by pointing out that "Canon City is distant from Denver something like 160 miles (257 kilometers) by rail. The Denver and Rio Grande Railroad passes southeast forty miles (65 kilometers), then turns north to Denver. This course is necessary on account of the range of mountains divided by the Arkansas River. This range consists of many lofty peaks in which Pike's Peak is included, almost directly in line between Canon City and Denver."

W. D. Harry of course had no idea what exact route the collie took (although the outward-bound trip was on that very railroad line), but really, the only surprising note in this story is that the dog returned to his former home when presumably he was living in the new home with his owner. In all the stories that follow, the point of the trip is to return to the new home and its familiar humans.

And there is no shortage of such accounts. There's "Bobbie the Wonder Dog" (also in the 1920s), who walked something like 4,000 kilometers (2,500 miles) to return home to Silverton, Oregon, after being lost in Indiana. It took the two-year-old dog six months, a journey that included crossing the Continental Divide.

Nor is it just dogs: In 2013, Holly the cat made it about 320 kilometers (200 miles) along the Atlantic coast of Florida back to her hometown. And while it's often hard to ascertain that these journeys actually took place, Holly had an implanted microchip, so it was certain she was indeed the cat she was thought to be. A cat in England traveled 217 kilometers (135 miles) to return to its old home, and Howie, a Persian cat, wandered 1,600 kilometers (1,000 miles) across the Australian outback to find his family. It took him twelve months.

While many such stories can be discounted on the basis of

misidentification of the animal, missing pieces of information, like a car ride, or plain mischief, the stories I've just quoted have been certified as real. But how are they possible? There is motivation: Cats are territorial enough to want to return to one they've been taken from, and dogs are attached enough to their humans to want the same. But the distances are baffling—even house cats that are free to roam seldom venture more than a couple of hundred meters from their home.

Aside from journeys undertaken by domestic dogs and cats, travels of huge distances in the wild happen as well, although some are extraordinary. Like this one: A young female Arctic fox wearing a radio collar headed west from the area of Spitsbergen, Norway, on March 26, 2018, and, a little more than two months later, settled down on Ellesmere Island in the Canadian Arctic. She had traveled 4,400 kilometers (more than 2,700 miles) across sea ice and glaciers, averaging a little more than 40 kilometers (20 miles) a day, but one day hit 155 kilometers (100 miles). Unfortunately, we have no idea if she might have wanted to return home—she is no longer being tracked.

There are return trips too. Bird migrations of a thousand kilometers aren't unusual, and even large mammals like caribou are capable of traveling large distances to arrive at a target location. These are natural migrations, undertaken annually or at least at some regular interval. But it's also true that many wild animals like deer, moose, bears, and wolves are capable of returning overland to a native territory from which they have been moved, like these rare examples of cats and dogs. It's common for "problem" bears that have been translocated at least 100 kilometers (62 miles) to return to their home territory in a few weeks.

Birds that migrate north in spring and return south in the fall, as well as homing pigeons, which can find their home loft from as many as 1,000 kilometers (620 miles) away, are known to use several different methods to guide their journey. They are able to track the earth's magnetic field (an innate ability) and the position of the constellations

Oddities

in the night sky (that's learned), and they can use one to calibrate the other.

Cross-referencing compasses is a crucial ability because, of course, cloudy skies render constellation watching useless, and even local geological magnetic anomalies can distort magnetic readings enough to bewilder homing pigeons, those champions of finding home. A sun compass is helpful too, but it must be linked to an internal clock—after all, the sun is due south only at noon, and if you don't know what time it is, you don't know where you're going. At least for migratory birds, there are other features that speak of home, such as local odors or landmarks.

In trying to figure out how your cat or dog could find its way home from, say, 100 kilometers (62 miles) away, it's difficult to know which—if any—of these multiple guides to long-distance travel would be in play. For instance, migratory birds need a few weeks to learn the movements of constellations through the night sky, but why would your dog or cat even have that potential ability? As you'll see in the next chapter, there is some evidence that dogs might have a magnetic sense to guide them over relatively short distances, but the experiments that have been done to establish that have also possibly involved other mechanisms, such as odor. Fine, but what odors could possibly be relevant over a very long-distance journey?

Consider where cats and dogs have come from. The ancestor of dogs, a now-extinct wolf, was a rambling, gambling hunter. Such wolves were definitely on the prowl over distance, and if modern wolves are any indication, they had territories. Modern wolves' territories wax and wane in size depending on the density of prey. When prey are relatively rare, wolf territories can be as large as 1,000 square kilometers (386 square miles). In Yellowstone National Park, they're much less: 427 square kilometers (165 square miles) on average. Thirty thousand years ago, when dog ancestors began to hang around human hunter-gatherers, their territories began to be shaped by those humans and inevitably would have shrunk. Although thirty thousand years is not a long time evolutionarily, the newly created dogs' genomes were being tuned to strengthen social ties to humans, not to maintain the ability to navigate. Cats, as we've already seen, are a very different story, and nowhere, at any time in the thousands of years of domestication of the house cat, have large territories played a significant role. Wolves pursued large mammals on the move; cats pursued mice. Even today, if a generalization can be made, cats are attached to their home, dogs to their humans. Of course there are house cats that ramble far and wide, but over a range where odors and landmarks likely play the major role.

A quick internet search reveals that most explanations offered for these impressive journeys invoke a magnetic sense, but the animals involved, Holly, Howie, and Bobbie the Wonder Dog, had no experience of the foreign location where they were stranded, magnetic or otherwise. It requires a heavy dose of faith to attribute their success to an internal compass. This is one of those pet attributes that at the moment we just can't explain—at least with science. Of course there are always those who think science is inadequate, and will turn to other—hypothetical—mechanisms to explain what seem to be extraordinary abilities (see chapter 25).

– CHAPTER 22 –

A Weird Relationship: The Origin of Dogs, Magnetism, and Poo

No account of the evolution of dogs would be complete without an acknowledgment of poop eating. My dog Robbie (unfortunately) engages in this habit whenever he gets a chance, which is seldom, if I have anything to do with it. I think it's disgusting, but I recognize that isn't really the scientific approach. Why does he do it? That simple question actually leads to a more profound one: Did poop eating play a role in the evolution of the dog?

So just to give Robbie his due, he doesn't eat his own poop—at least now. I'm pretty sure he did when he was a puppy. But he seems to delight in other dogs' poop, and while he rarely encounters the fecal matter from other species, like horses or cows, I'm pretty sure both would interest him. Fascinate him, even. For the record, I've read all the supposed reasons he does this, including inadequate nutrition, boredom, parasitism, mineral deficiency, vitamin deficiency, even copying other dogs, and I can assure you none apply. I'm also fully aware of the study by Benjamin Hart and his team at the University of California, Davis, which revealed *not one* of eleven surveyed treatments for curbing the ingestion of feces actually worked.[1] Although, to be honest, I haven't

A Weird Relationship: The Origin of Dogs, Magnetism, and Poo

yet followed the advice of the woman I met on a dog walk who said I should feed him bananas.

Why bring this up in connection with the evolution of the dog? Because there's important data that sheds light on the practice. First, any time there are parasites among a wolf pack, it benefits the wolves to consume parasite-laden feces within the first two days before the eggs within the feces hatch into infectious larvae. That's not precisely a factor in the transition from wolves to dogs, but it does at least give such behavior a rationale. But with respect to the wolf-to-dog transition—and here we're getting very far into the speculative realm—there is evidence from all over the world that free-ranging dogs, especially in warm climates, eat human feces routinely. Now turn the clock back thirty thousand years and picture wolves hanging around small clans of hunter-gatherers—what is there always an abundance of? That's right.

So I've come to the conclusion that rather than engaging in a disgusting habit with truly off-putting relish, Robbie is simply hearkening back to a crucial step in the evolution of his species. It makes me feel better. Briefly.

Then there's production, the partner of consumption.

Robbie is also unusually diligent in his pre-poop spin, sometimes doing four or even five complete circles before settling in. Of course he's not alone in spinning; most dogs do it—but why? As with any pet behavior for which the reasons aren't immediately obvious, there's much speculation and little evidence. Maybe the dog is checking the surroundings for potentially hostile other dogs, or people, or whatever. I don't think Robbie is doing that because he appears to be staring at the ground immediately in front of him. You get the feeling that he's concentrating on something that has nothing to do with his immediate surroundings, like obeying some atavistic instinct.

Some have suggested that's exactly it: Maybe spinning dogs are leaving their scent on the ground under them with their feet (you'd think the poop would need no help in that regard), or perhaps they're just smoothing down the turf—especially if it's long grass—but again, there's not really any good evidence for any of this. On the other hand, dogs spinning before lying down to sleep makes good intuitive sense at least—leveling out the surface for greater comfort.

However, one explanation for spinning does have some experimental support, though it's controversial. In 2013, a research team from the Czech Republic and Germany analyzed the tendencies of both female and male dogs to point in a preferred direction—any direction—while urinating and defecating. After collecting a magnificent total of 7,475 observations (1,893 defecations and 5,582 urinations) from seventy dogs of thirty-seven different breeds, they found—in a bizarre connection to the voluminous data on dogs being able to find their way home—that both sexes preferred to have their spines aligned with the north–south magnetic axis. Not all the time, however—only when the magnetic field was calm, because it is normally in some sort of random flux one way or the other.

There has been follow-up on this initial study, but given that it's been both positive and negative, it hasn't exactly clarified the somewhat mysterious nature of these findings. On the positive side, the same Czech-German research team expanded this original study to hunting dogs and their ability to find their way home. Twenty-seven different dogs were released at a variety of locations in the woods, more than six hundred trials in all, and the research showed that the dogs used two different strategies for their return journey. One was to simply follow their outward-bound route in reverse (dogs used this approach in nearly two-thirds of the trials), but in the other, more mysterious version called scouting, dogs chose different, often shorter routes back to where they started.

What made the scouting version intriguing was that a dog would begin by running a short distance, around twenty meters (65 feet), in a north–south orientation, whether or not that actually took them closer

A Weird Relationship: The Origin of Dogs, Magnetism, and Poo

to home. Dogs that did that initial run were more efficient in getting home. The researchers suggest that dogs that did that were somehow resetting a kind of mental map and calculating their route to coincide with north to south.

Note this experiment only suggests that dogs are sensitive to magnetic field lines, and says nothing about why they should prefer to line themselves up with those lines when excreting. So it's definitely not a replication of the original study, and indeed, there has been pushback.[2] In 2021, a pair of scientists at the University of St. Andrews, Anna Rouviere and Graeme Ruxton, searched in vain for evidence of dogs using an internal compass to "do their business" (my favorite euphemism). They tracked the "activities" (another euphemism) of dogs in five dog parks in Lyon, France. They set their experiments in dog parks to align their results with previous dog park experiments that found evidence of a magnetic effect. However, their fifty-two dogs exhibited no detectable magnetic tendency.

I have to add here that in a slightly less rigorously scientific test of this theory, we did our own version for *Daily Planet* at the Ontario SPCA. In what looked to be—on camera, at least—a carefully orchestrated demo involving four different dogs, of different breeds, we brought them to outside pens in a snowy yard and within seconds they chose to defecate. I have no idea if they were somehow prepped for it, but they were certainly much more predictable than my dog. The results demonstrate the difference between science on TV and real science. Of the four dogs, two lined up (roughly) with the north–south axis, one didn't, and one was in the act of turning as it happened, so that result had to be thrown out (so to speak). I nonetheless proclaimed the experiment as support for the theory.

Where we stand at the moment, then, is that, as we've already seen, it's not particularly surprising that dogs would be sensitive to the earth's magnetic field. But why should that have anything to do with where to stand when peeing or pooing? I'm cautiously agreeable with

Oddities

the idea that spinning before settling down might be a way of checking out the surroundings (although, as I say, I don't see my dog doing that at all), but how the magnetic field plays into that is mysterious. Some have suggested that defecation is a territorial marker. Maybe so, but should it matter that the dog stands in a particular direction? After all, it's the odor, not the angle, that matters.

– CHAPTER 23 –

Do Dogs Know Calculus?

"Do Dogs Know Calculus?" is the title of a paper in the May 2003 issue of the *College Mathematics Journal*, by Timothy J. Pennings, a math professor. His question is serious, at least to the extent that he performed experiments with his dog Elvis, although his use of the word "know" might be a little dubious. I'm a big fan of this paper, not only because it's ingenious, and does follow scientific protocol, but also because I replicated it with my dog Buster (RIP), admittedly in a less scholarly fashion.

Pennings wanted to explain the "optimal path problem"; that is, how to calculate the fastest route between two points, A and B, when the journey is complicated by the fact that you must travel through two different media, such as over solid ground, then through water. The example he picked, recruiting his corgi Elvis, was to stand on a beach and throw Elvis's favorite tennis ball into the water.

Elvis's goal is to retrieve the ball as quickly as possible (Pennings drew that conclusion from observing Elvis's obvious excitement). Depending on where the ball lands, that accomplishment is either straightforward or somewhat complicated. Throwing the ball straight out into the water simply requires Elvis to swim straight out to it—no problem. But throwing the ball at an angle, right or left, opens up a choice of options. Assuming Elvis can run faster than he can swim, what is he to do? He can jump immediately into the water and follow the flight of the ball directly, but that means spending a lot of time in the water at a relatively slow speed. Alternatively, he could run along the beach until he is directly in front of the ball and then swim straight to it. That's not a bad option, but the total distance traveled is greater than if Elvis had swum out at an angle.

There's a third alternative, though, which is to take the best of both alternatives: run along the beach a certain distance, then swim at an angle toward the ball. If done exactly right, that approach will maximize the value of speedy running along the beach and the shorter distance of swimming. But where exactly to take the plunge? If you're a human trying to solve this problem, you'd resort to calculus. But if you're a dog?

Pennings observed that Elvis *seemed* to be following the third strategy, but of course "seemed" isn't nearly good enough. So he embarked on an elaborate set of measurements and calculations, timing Elvis's speed on land and in the water, measuring exactly where along the beach Elvis stopped running and started swimming, then comparing that location with where the calculus had demonstrated would be the ideal place. Et voilà! The dog was a mathematician. Or at least close enough.

Pennings pointed out that there were some simplifying assumptions he'd made that a skeptic might be quick to point out. He assumed that Elvis started swimming as soon as he entered the water, but he could have waded some distance first (although given that he was a corgi, he couldn't have walked very far); in addition, because of wave action, the ball wasn't stationary in the water. Also, the waves meant there was never a hard and fast line marking the edge of the beach and the start of the water. But really, all of these seemed pretty minor. Pennings also admitted that he "omitted the couple of times when Elvis, in his haste and excitement, jumped immediately into the water and swam the entire distance. I figured that even an 'A' student can have a bad day," he said.

Do Dogs Know Calculus?

I found this study inspiring, to say the least. At the time I read it, Buster and I were working on a TV science show called *Daily Planet*. Doing the TV version just seemed like an obvious thing to do, so we did.

In any measurements of dog intelligence I've seen, poodles outrank corgis—"outstanding" versus "very smart" seems like a typical comparison—so naturally, I expected that Buster would do what Elvis had done. I should have realized it's never wise to make assumptions when it comes to dog behavior.

Up to a point, everything went according to plan. I threw the ball at the appropriate angle from where we were standing. Buster raced down the beach all right, but a little farther than I had expected—in fact, all the way to the point where the ball was bobbing up and down directly in front of him. Then he walked out toward the ball, covering a much greater distance in the water, given his long legs, than Elvis could have. Then he stopped. And waited for the ball to be carried toward him by the waves. Actually, he returned to the beach and waited.

I said at the time, "Now that's intelligence!" Of course there are other possible factors. Buster's motivation? Did he not like the cold waters of Lake Ontario? Was he really not that jazzed about chasing an old tennis ball into the water? Fortunately, this was a TV show, not a calculus lesson, and the truth is, that medium allows you to make something out of *any* conclusion to an experiment. And so we did. As far as we could see, he was just too smart to bother.

– CHAPTER 24 –

Do You Look Like Your Dog?

A simple question for all of you who share your home with a dog: Do you resemble your dog? I mean facially, of course! If you and your dog stand in front of the mirror, or you take a cheek-to-cheek selfie, are you recognizably similar in appearance? It sounds absurd—canids and primates have been evolving separately for sixty million years or so—and yet . . .

In 1999, Professor Stanley Coren of the University of British Columbia published a scientific paper called "Do People Look Like Their Dogs?"[1] Coren wasn't the first to suggest that dogs and their humans look alike—both the cell phone company Fido and the dog food company Cesar ("He/She might look like you. But he/she doesn't have to eat the same food") created ads showing stunningly similar humans and dogs. The likenesses were so impressive it seemed highly unlikely the humans and dogs were partners in real life, but the ads were memorable nonetheless. But Stan Coren chose to put the idea to the test.

Coren first made the case that there was some psychological data that hinted the dog-human similarity might be true, one of which was the finding that humans tend to select mates based on similarity of appearance to themselves. You'd be justified in thinking this is a step too far, especially if you're not partial to dogs, but Coren argued that this could be "carried over to selection of a canine companion." It hearkens back to the idea that baby-faced pets take advantage of the

Do You Look Like Your Dog?

human tendency to be attracted to baby faces. Definitely not evidence, but stage setting (more of this later).

We are also *very* familiar with our own faces—especially the left-right-reversed version we see in the mirror.* But how to establish there might be some truth in this commonly held but thus far unsubstantiated belief? Coren decided to test the idea that humans might be attracted to dogs that look somewhat like them.

His method was straightforward enough: 261 female students at the University of British Columbia were asked to rate images of four different dog breeds for how much they liked the dog's look, how friendly it appeared, how loyal it might be, and how intelligent, features that together could be aligned with wanting to associate with the animal.

The breeds (headshots only) were a springer spaniel, a beagle, a Siberian husky, and a basenji. Two notes about the experimental setup: Coren chose women students because he had hypothesized that hairstyles might predispose the wearer to certain canine looks, especially those that imitated the outline of hair. For instance, long over-the-ears hair versus a short cut exposing the ears might play a critical role in preferences for floppy dog ears versus erect, pointy ears. (Male students weren't invited because their haircuts were just too uniform.) Coren also asked the participants to review sketches of hairstyles and choose which was closest to the way they usually wore their hair.

The results were consistent with what he had apparently expected: Women with long hairstyles judged dogs with floppy ears (the beagle and the spaniel) as being likeable, friendly, loyal, and intelligent. Women with shorter hair gave the higher ratings to the husky and the basenji.

Coren suggested that the relation of hairstyle (humans) to floppy or prick ears (dogs) is really only one criterion, and others could be tested,

---

* Is it weird that there's some evidence that we prefer the reverse mirror image of our face to the unreversed version that everyone else sees?

like slim versus robust body build or "nasal prominence" (noses were a major feature in some of the commercials mentioned above). One obvious shortcoming of this experiment was that it had nothing to say about dog ownership, just appearance, and you could easily imagine that countless other factors could influence the actual acquisition of a dog.

But the Coren experiment was only the beginning. Next up were Nicholas Christenfeld and Michael Roy, whose report "Do Dogs Resemble Their Owners?" was published in the journal *Psychological Science* in 2004.[2] Starting with pictures of forty-five dogs and their owners, the researchers presented volunteers with three pictures: a dog owner, that person's dog, and a second, random dog. No one involved knew anything about the dogs or their owners, but were simply asked to choose the dog they thought belonged to the human. Any owner in the study appeared over and over, with his dog and an ever-changing companion dog. Christenfeld and Roy took great care to ensure there were no hints as to which dog belonged with the human and which didn't—for instance, no evidence of the dog park background where the photos were taken was apparent.

The results were unequivocal: Owners were more likely to be paired with their dog than with the alternate—but *only* if their dog was a purebred. Owners of non-purebred dogs were only linked with their dog at a chance level. This makes sense because there's a reasonably good chance to be able to predict what a purebred puppy will look like as an adult, compared with one whose genetics are unknown. Also emerging from the study was the fact that the duration of ownership was not a significant factor in resemblance. This makes it doubtful, as some have suggested, that a human and his/her dog might grow to resemble each other over time, if only by being clipped in a similar style.

On that note, however, Stanley Coren's suggestion—that the shape of the hair framing the face matched a dog's ears—didn't seem to play a role. Long or short hairstyles didn't match with lop or prick ears. Nor did female or male ownership, or "sharpness of features, attractiveness, perceived friendliness or perceived energy," leaving open the question of what exactly accounts for the ability of independent judges to perform better than chance at linking the owner of the dog with the dog.[3]

And here is where the authors end their study with a reference to an idea that Stan Coren had briefly mentioned: "It does appear that, as in the case of selecting a spouse, people want a creature like themselves."

The idea that couples resemble each other has a long history (it's called "assortative mating") and is based on many more attributes than simply appearance. Appearance counts, but so do religion, socioeconomic status, age, personality traits, neighborhood, social attitudes, and more. There has even been research showing that people with the same blood type tend to gravitate toward each other, a fascinating finding given that few of us ever exchange details about our blood type (but could it be detectable in some as yet undiscovered way?).

It does seem to be a stretch to attribute this similarity between dog owners and their dogs to assortative mating, even given the well-known tendency of dog owners to refer to their pet as "part of the family." At the risk of this spiraling out of control, I have to leave you with two more studies.

A recent review of this research added a curious piece: The study reinforced the idea that owners do indeed look like their dogs, but also share personality traits, based on the Big Five mentioned in chapter 14 (neuroticism, agreeableness, openness, conscientiousness, and extraversion). The authors were hesitant to conclude why this is, suggesting either that humans may choose dogs that mirror their psychology, or that complex interactions between owner and dog over time tend to align their behaviors.[4]

And finally: Two Viennese psychologists, Stefan Stieger and Martin Voracek, added their two cents to the resemblance phenomenon: They included *cars* as well.[5] Again, participants were able to match car owners to front views of their cars (versus other cars). They went to great lengths to neutralize the effect of car stereotypes, like more men than women driving off-road vehicles, and then went one crucial step further and showed that the owners' cars resembled the owners' dogs. Only, however, if those dogs were purebreds.

As I was writing this, I owned a 2014 silver Mazda 3 and was custodian of a five-year-old poodle–wheaten terrier cross. Somehow that doesn't help me understand what's going on here. However, if you are skeptical that the front of a car can look like a human's or a dog's face, remember the launch in 1989 of the Mazda Miata, that cute little roadster? My thanks to CarAndDriver.com for this description: "Its features are endearing, including pop-up headlights that, when opened, give the MX-5 the appearance of a happy, wide-eyed thing—especially in conjunction with the rounded 'mouth' in the bumper."

– CHAPTER 25 –

A Psychic Dog

When I started writing this book, I didn't expect I'd have to include a chapter on how pets became tools in an ideological conflict between those who are skeptical of science and those who are skeptical of those people. But I did. In this case, Rupert Sheldrake is the one who promotes the idea that the world is replete with forces unacknowledged by science, and Richard Wiseman is the one who claims there's no evidence for such forces, especially those Sheldrake promotes. Oh, and there's Jaytee the dog.

Sheldrake landed the first blow in this back-and-forth in the mid-1990s, when he publicized the TV coverage of a set of experiments monitoring the dog Jaytee, a terrier mix owned by a Manchester woman named Pamela Smart. Smart felt Jaytee demonstrated telepathy because he would hang around the front door or living room window when Pam was on her way home, and, crucially, he was seemingly aware she was on her way while she was still too far away to be seen or have the unique sound of her car motor recognized. Sheldrake set up an elaborate plan to determine whether this apparent mind reading was actually happening. He recognized that there were several alternative (and humdrum) explanations that an experimental setup would need to eliminate that might not have been accounted for in the TV coverage.

These included ensuring that Pam's setting out for home happened at random times so that Jaytee couldn't settle into a rhythm based on time of day; that Jaytee didn't routinely go to the living room window every five minutes or so; that, as Pam asserted, Jaytee would have to be on alert for her arrival long before the dog could either see or hear her returning; and that no other people in the home cued the dog by some behavior of their own based on their knowing that Pam was on her way.

The precise details are available online.[1] In general, though, Sheldrake settled on having Pam leave home and stay away until cued to begin to return by choosing times with a random number generator. The cue was often delivered to her by pager from Sheldrake, who was three hundred kilometers (186 miles) away in London. Meanwhile, Jaytee's behavior was being monitored nonstop by a time-stamped video camera.*

A sample graph of the results presented by Sheldrake shows three ten-minute periods. The first covers a time when nothing remarkable is happening or is about to happen. The second is the ten-minute period just before Pam receives the pager signal to start the homeward journey, and the last ten minutes covers the first part of the drive home. (Usually the drive home is significantly longer than ten minutes, excluding the possibility that Jaytee could hear or even see Pam's car.)

Sheldrake's analysis showed that Jaytee visited the front door or window only about 4 percent of the time in the first ten minutes, 35 percent of the time over the next ten, and 55 percent during the final ten. The most straightforward interpretation of these results is that he definitely seemed aware that Pam's arrival at home was imminent.

It's no surprise that Rupert Sheldrake interpreted these results as evidence that Jaytee's mind was somehow telepathically connected to Pam Smart's mind—no surprise because Sheldrake's career has been

---

* I know, I know: pagers and video cameras—it was the twentieth century!

A Psychic Dog

devoted to the idea that modern science is not open to phenomena that could roughly be lumped under the label "paranormal," and when faced with results like these, he would find an explanation for them that would run against scientific orthodoxy. Telepathy is definitely one of those. There was little chance he would scan the data, shrug his shoulders, and admit there was no conceivable explanation (to be fair, few scientists would).

Besides, Sheldrake already has an explanation—his own, called morphic resonance. This is his theory that there are fields connecting members of a species that allow them to gradually build group memories. A favorite example of Sheldrake's is that if laboratory rats are continually run through a certain kind of maze, with time new generations of rats will learn the maze faster, because through morphic fields, they will be absorbing the learning of their fellow rats, even of previous generations.

In the case of dogs anticipating their owner's return—and Sheldrake has also published similar results with a Rhodesian ridgeback named Kane—the dog is tapping into the morphic field of the owner. In Sheldrake's own words: "The morphic field hypothesis predicts that organisms that have been part of the same social system from the past, part of a bonded group, will remain connected at a distance."[2]

All right, then—case closed. Actually, not quite. At some point in all this excitement about Jaytee in the mid-1990s, Richard Wiseman, a psychologist and skeptic, and his colleagues did their own, videotaped, double-blind, random-number-generated version of the Sheldrake experiments and came to—wait for it—the opposite conclusion. They didn't collect as much data, and they devised a slightly different design for their experiment, but in tracking Jaytee's trips to and from the front door and living room window of Pam Smart's home, they found nothing to convince them that the dog was timing his visits to coincide with her imminent arrival. Again, you can check out their account online.[3]

Oddities

The frustrating thing about the two opposing views of what the data shows is that the data collected both by Wiseman and Sheldrake are not dramatically different. If anything, the boundaries between what each deems as acceptable and not acceptable data seem to be the critical factor. Sheldrake argues that their data are essentially the same. Wiseman, in a podcast interview, argued, "When I look at [Rupert's] data, there may well be something going on, they don't look to me quite as methodically rigorous as you would need in order to be able that make that decision firmly in one direction or another."[4]

Having earlier labeled Sheldrake as someone who "promotes the idea that the world is replete with forces unacknowledged by science," it's only fair that I repeat my description of Wiseman as someone who claims "there's no evidence for those." It's therefore not surprising that where Sheldrake sees a fascinating example of animal-human telepathy, Wiseman sees inadequate data. It's frustrating to those of us who would have liked to see a full-on, detailed, perfectly controlled experiment that might have provided some sort of concrete result, but that didn't happen, and given the time and expenses involved in experiments like this, it's not likely to.

Does it matter? This isn't dissimilar to the examples of long-distance trips by pets to rejoin their owners that I described in chapter 21. To me, there are no definitive explanations of how dogs and cats can do that, magnetic field perceptions aside. Having a magnetic sense is one thing, but unless you can connect it to some sort of mental map of where you're going, a trip of hundreds of kilometers just isn't going to happen. Rupert Sheldrake would likely attribute those journeys to shared thoughts via morphic resonance, but there are no data to support that. The difference in that case is that there actually *is* evidence that some animals have managed to do that, so skeptics like Wiseman aren't able to deny that it happens. I haven't yet seen a skeptic address those "impossible journeys."

This is a tempest in a teapot when it comes to the world of science,

A Psychic Dog

and in fact, there's another area of activity that has totally shrugged off the controversy, and that's the world of pet psychics. Everywhere you turn, there are pet psychics who, for anywhere from $250 to $500 per half hour, will help you connect with your pet. How do they do that? My favorite example is Nikki Vasconez, a former lawyer in Pennsylvania who dumped her law practice as she gained self-taught expertise in communicating with pets telepathically. Her practice does strike you as the ultimate in convenience: She doesn't even have to actually meet your pet; a photograph will do. As she told the *Daily Mail*, "I'll know the animal's name, gender, and whether they're living or in spirit. Then I look at the animal's photo and I take a couple seconds to picture my heart connecting with their heart—I'm a very visual person so I imagine a beam or a light connecting the animal and I through our hearts."[5]

Ninety minutes for $550. There's a waiting list.

– PART VII –

Thinking Long-Term

– CHAPTER 26 –

The Best-Before Date

It's common to advise a potential buyer of a parrot that "it might outlive you." The warning jolts: Pets just don't outlive people. Usually they don't even come close; if you're over thirty, every dog that was alive when you were born is now dead.

The death of a pet is a difficult event. Humans grieve. Sometimes over and over. It never seems to get easy. So naturally, when facing such a moment, many people wish they had done something—anything—to extend their pet's life.

You can do your best to ensure that your dog or cat (or whatever) is fit, healthy, and has a good life. But the influence of all that is constrained by the maximum lifespan of the species. Actually *extending* the life of pets is not yet possible, but is an ongoing research effort. The Dog Aging Project in the United States aims to create longer-lived pets; specifically, in this case, dogs (details in chapter 33). Researchers are giving dogs rapamycin, a compound that has shown promising life-extending effects in rodents, but this clinical trial is still underway, so it will be a while before there's long-term data out of it.

Outside of rapamycin, there's no life-extending drug for dogs or cats or any other pet in existence. Until there is—unless you have oodles of time to wait until selective breeding gives you a dog that lives to twenty—you still have multiple options. One of them yields a pet that is as close to the original as possible, one keeps your pet in a suspended state, and two others make it possible to keep your pet around with you forever, with one caveat: It won't be alive.

The choices among these approaches depend partly on how much cost and risk a pet owner is willing to bear, and how much forethought has gone into the decision. Let's begin with the most common case: The pet owner really hasn't even thought about the state of the pet until *after* it dies. Many take a few days before realizing that's it! Little Truffles is gone.

It's at this point that people begin to assemble photos in some kind of pet album, maybe bury their pet in the backyard, have an impression made of a paw, or have it cremated. But there are also those who want the animal to be around *exactly as it was*. There are two ways of trying to do that, freeze-drying and taxidermy, but even these need some forethought before the actual death. In either case, the body should be put in a freezer as soon as possible, since both methods require a body that has not begun to decay. An hour or two might not be too damaging, but an owner who wants to preserve the pet forever would surely want to start the preservation with the pet exactly as it had been in real life.

Freeze-drying a pet is in some ways the most direct way of creating a long-lived version. After thawing the pet from the original freezing, the pet owners consult with the freeze-dryer as to what permanent pose they'd like their animal to have. Once posed, the long process of refreezing and drying begins. The goal is to bring the corpse down to temperatures where the ice in the body very gradually changes directly from solid to gas, ice to water vapor. Over months, all the water that had been in the corpse has now escaped, and because there was no movement of liquid water, the shape of the animal and its internal integrity has remained constant. With the exception of artificial eyes, and some internal organs having to be replaced by packing material (because those organs don't dry well), Little Truffles is back, sitting there in his trademark pose. Forever.

The Best-Before Date

Freeze-drying a pet isn't cheap, nor is it easy to have done with a bigger animal, like a thirty-five-kilo (seventy-seven pound) dog, but one advantage is being able to create the exact pose one wants.

Taxidermy is quite different, in that the skin, the pelt of the animal including paws, and the teeth are the only original parts retained. The interior of the animal—bones, organs, muscles—are discarded, and the skin is draped over and wrapped around a mount. The results can be absolutely amazing. The best taxidermists are dedicated to making the animal they're working on, whatever it is, as lifelike and expressive as possible.

Taxidermists too consult extensively with owners before they'll agree to work on their pet. Obviously, it's an emotional issue. One of the differences here is that taxidermists, most of them anyway, are artists, and like to have a creative hand. But when working on pets, they're constrained by the owner's desire that the animal look exactly as it used to. Those owners are generally not interested in a dramatic pose that they never saw their pet adopt. Of course, with freeze-drying, that tension is absent. The freeze-dryer isn't wanting to apply his creative flair to the pet, just dry it.

Allis Markham is a preeminent American taxidermist in Los Angeles. She is a much-decorated artist who creates specimens for museums, does commissions, and was part of a seriously good documentary called *Stuffed*. (Of course animals aren't really stuffed at all.)

Her company, Prey Taxidermy, does pet taxidermy, but she posts a list of cautions on its website for anyone contemplating having a pet taxidermied.[1] It's pricey ($6,000–$7,000), and there are limits to the variety of poses possible (mammals will, most of the time, be sleeping with eyes closed). But the concern, both for the client and Allis herself, is disappointment: "[The] biggest challenge is fear of failure. Such a weight to it. The person has stared at this pet all day, every day for how many years? Then, having never met this animal alive, the taxidermist is meant to make it *look* alive, and capture the little expressions

Thinking Long-Term

and personalities that domestic animals have literally evolved to have, without ever having met it alive to know its personality and see those expressions. You get one shot."[2]

There are challenges, to be sure: Overweight, elderly pets have rolls of fat that create issues of fit, while aged pet faces have lost some of their fur. I'm sure that a taxidermist like Allis Markham will produce a work of art. Still risky, though: "If the outcome isn't what they imagined, that can be very devastating—they've lost their pet twice."

On the one hand, I'm not surprised at the number of people who want to use one of these methods to preserve their pet after death. Nor would I be surprised if many of those same people would say that this is good evidence that their pet was a valued member of the family, even equivalent to the humans. Equivalent? Not quite: No other members of the family are freeze-dried or taxidermied.

– CHAPTER 27 –

Bring 'Em Back Alive

Freeze-drying and taxidermy have a strong appeal for those who can't imagine living without some sort of presence of their deceased pet with them. Others, dissatisfied with this approach, want a living version of their pet. The two ways of keeping it alive beyond its years are both extremely tricky, each at this point a Hail Mary at best, and neither exactly extends the life of the pet. One, in fact, is so unlikely as to barely deserve mention. Nonetheless, here we go:

"Insights from One Thousand Cloned Dogs" is the title of a recent paper in the journal *Scientific Reports*. It is just what it says: a summary of what has been learned from cloning a thousand dogs over the last two decades. (It was published in 2022, and the lag between research and publication means that the number of cloned dogs is actually closing in on two thousand today.)[1]

You had no idea that dogs were being cloned, let alone a thousand of them? Or that any animal had been cloned? Cloning has become an industry, and while the cloning of pets is a small part, it raises issues that come closest to us all.

Cloning has a strange history. Its scientific lineage is straightforward. In the early 1960s, British scientist John Gurdon managed to create clones of frogs, for which he received the Nobel Prize in 2012. In typically opaque fashion, the Nobel committee explained that he had won "for the discovery that mature cells can be reprogrammed to become pluripotent."

What Gurdon had shown was that even though cells in the body specialize to become liver cells, neurons, and skin cells (something like two hundred different kinds in the human body), each of those cells still possesses all the genetic material necessary to start over, to produce a fully formed adult version of whatever animal—in Gurdon's case the African clawed frog, a lab favorite.

The technique sounds simple: Remove the nucleus (containing the genetic material) from a frog's egg and replace it with the nucleus from an adult cell, in this case from the intestine. That new, genetically identical tadpoles could develop from this engineered egg showed that intestinal cells had retained all the genes necessary to generate an adult frog.

Important as that demonstration was, frogs are relatively easy because once the eggs are laid, they're on their own. Mammals are much more difficult. Replacing the nucleus from an adult cell is the same, but it then has to be transplanted into a surrogate female whose reproductive system is prepped to bring that egg to the embryo stage. By now, many mammals, including sheep, rats, buffalo, mules, camels, goats, deer, and macaque monkeys, have been cloned. Dolly the sheep was the first, in 1996. But long before that, the pop culture world seized on cloning as the platform for spectacularly unsettling ideas—like the cloning of humans.

It started in 1976 with a novel by Ira Levin called *The Boys from Brazil*. The plot, in its stripped-down form, had the evil Nazi Josef Mengele cloning Hitler and using his tissue to create ninety-five clones, whom he hoped would create the Fourth Reich. The book was turned into a movie, which received mixed reviews, despite an all-star cast (Laurence Olivier, James Mason, Gregory Peck) and three Academy Award nominations.

The Boys from Brazil was admittedly fiction, but *In His Image: The*

Bring 'Em Back Alive

Cloning of a Man, a 1978 book by David Rorvik, a widely published science writer, claimed to be a true account of the birth of a cloned boy. (Rorvik's book is not to be confused with *In His Image*, a novel by James BeauSeigneur about Jesus Christ being cloned from cells found on the Shroud of Turin.) Rorvik was tight lipped about who had actually performed the cloning, where the offspring was, and anything else about the event that could be investigated. He did say that, in 1973, he was approached by a businessman named Max who wanted to be cloned. Rorvik said he participated in setting up the preparatory experiments, and eventually a clone of Max was born to a surrogate named "Sparrow." In hindsight—and even at the time—almost no one with any knowledge of cloning believed it. Today? It's forgotten.

But this brief flurry of fascination with cloning left its mark; cloning has ever since had this eerie mix of threat and promise. There is no evidence today that a human anywhere has been cloned, but for other animals, the technique has become widespread. Cloning has produced teams of polo horses, cows that produce more milk, and pigs that are meatier. Chinese scientists have even reportedly used robots to produce cloned pigs more efficiently. And then there are pets.

How did we get to "one thousand cloned dogs"? Ironically, it started in 2001 with a cat named CC (Carbon Copy or Copy Cat, whichever you prefer). Success in this initial case was hard-won: Eighty-seven cloned embryos were implanted into eight surrogate female cats, resulting in a grand total of one failed pregnancy and one success. While the odds have improved since then, cloning is rarely straightforward.

The first dog(s) were cloned in 2005 in Seoul, South Korea. Of the two born, one, an Afghan called "Snuppy" (Seoul National University puppy), lived for ten years. The other unnamed puppy died after only ten days. Snuppy himself was cloned, and three out of four offspring survived.

Snuppy the dog was the fifteenth species of animal to be cloned. The experience gained, or the nature of dog reproduction itself, has shown that while cloning dogs seems more efficient than cloning many other species, there are still technical details to be worked out. Many of them revolve around timing: When is the surrogate mother hormonally ready for transplanted embryos? When are the embryos ready? How long can they remain outside their incubator or the mother?

Dogs aren't unique—every species poses unique biological problems for cloning. But fine-tuning these issues, at least with dogs, has resulted in greater success. In contrast to the rate for CC the cat of eighty-seven transplanted embryos with one resulting kitten, one Chihuahua generated nineteen puppies in thirty-two surrogates. The only more efficient donor on record was a coyote whose cloned eggs were incubated by female dogs.

So dog cloning is solidly established. But why do it? One application has nothing to do with pets. Dogs share something like 350 genetic defects with humans, and from the medical research point of view, being able to clone such dogs would expedite the research into these defects.

But there is also a consumer demand for cloning pet dogs. Barbra Streisand, already famous, became more so with the revelation that she had had her dog Sammie cloned after death. Sammie was a curly-haired (they're commonly straight-haired) Coton de Tuléar, and Streisand really wanted another curly-haired version. But because they're rare, she opted for cloning. She forwarded cells that had been taken from Sammie before the dog's death to ViaGen, an American company specializing in pet cloning. Fifty thousand dollars later, Sammie had been successfully cloned into four puppies. One died, Streisand gave away two, and she kept one. Paris Hilton has also had a dog cloned.

There are at least three issues with having your dog or cat cloned. In no particular order they are misplaced expectations, the ethical issues of the process, and the missed opportunity to adopt one of the hundreds of thousands of shelter dogs.

The genetic identity of the cloned pet gives rise to the expectation that it will be just like the deceased version, but that is unlikely for two reasons. One, there is more to genetic expression than simply the DNA you're born with. There are mechanisms in cells that alter or influence the expression of the genes. Those influences in the pet whose DNA is the source of the clone will not be the same as the influences in the cloned animal. This was clear from the beginning with the first cloned cat. CC was in every respect a successful clone, living to eighteen years of age, but her multicolored coat was quite different from the cat she was cloned from. It was already known that coat color in such animals is determined by factors in development, not controlled completely by genes.

Coat color might not be an issue for a pet owner, but what about temperament and behavior? These too are not predictable; the experiences of a pet as it matures and its environment help shape its behavior, and it is highly unlikely the cloned pet's experiences are exactly the same as its progenitor. To be fair, many cloned pets do look like replicas, and owners of cloned pets see behaviors that remind them of their previous pet, but expectations that the clone will inevitably be exactly the same are unrealistic.

The ethical issues revolve around the process of cloning. Other, unnamed dogs are recruited to make cloning happen. First, a female dog "donates" eggs to serve as the vehicles for the pet-to-be-cloned's DNA. This includes hormonal stimulation and then surgery to remove the eggs. A second dog must then serve as a surrogate to incubate the eggs and give birth to the puppies (or kittens). Jessica Pierce, an ethicist, has called these dogs the "underclass"—they're used to satisfy a human's desire without any benefit to themselves.[2] What then happens to these unseen participants?

As Alexandra Horowitz, the author of *Inside of a Dog* and a scientist at Barnard College, explained in a *New York Times* article on dog cloning, ViaGen is unwilling to reveal much about the egg donor and surrogate dogs, other than to say that they rent them, then return them to their "production partners" who then may use them for "other projects." The equivalent South Korean pet cloning group stated in the "Insights from One Thousand Cloned Dogs" publication that they use Korean mixed-breed dogs, both for egg recovery and surrogacy, and "animals are used only once for a cloning event and well cared for, with the consistent aim of decreasing animal use and reducing any potential stress or suffering."[3]

If concern over the egg donors and surrogates and their eventual fates isn't enough, then it's worth considering the choice: Cloned dog/cat or rescue? The best estimates are that more than six million dogs and cats are admitted to shelters in the United States every year. Roughly half are dogs, half cats, and about 370,000 dogs and 300,000 cats are euthanized annually. Despite an uptick in adopting shelter animals, those numbers are still staggering.

So then, if money is no object, acquiring a new cat or dog offers choices: purebred from a breeder, rescue or adoption, or cloning. The motivation for cloning is the urge to continue, in some slightly altered form, the relationship that has already been established with a pet. The plus is that your life with that pet is dramatically extended—except there's no guarantee it will be "that pet." There will be differences, although perception bias—the tendency to persuade oneself that the clone is behaving much like its predecessor (because we expect that)—must be operating. However, many owners of cloned pets testify they are happy with the results. But to be completely happy, they must be relatively unconcerned about the involvement of the other animals involved or be willing to spurn adoption.

As far as motives go, Barbra Streisand's was not unusual: "I was so devastated by the loss of my dear Samantha, after 14 years together, that I just wanted to keep her with me in some way. It was easier to let Sammie go if I knew I could keep some part of her alive, something that came from her DNA."[4] To her credit, her love of dogs was as powerful as her love of *that* dog: While waiting for the lab results, she adopted a shelter dog.

Bring 'Em Back Alive

Cloning is scientifically fascinating, but cloning pets has an ethical cloud hanging over it. It's worth remembering that while cloning your pet, if it works out, gives you extra years with a loved pet that is probably much like the original, acquiring a different dog allows you a chance to build a different relationship with a different animal.

There is another way you could extend your dog's life without the risk that a clone turns out to be not quite what you expect: cryonics. This is the technique of freezing your pet after death and maintaining it in that frozen state until humankind has figured out how to bring it back to life. Which could take centuries.

Cryonics comes with an important disclaimer: It's never been done. But rather than sound like one of those naysayers who are eventually—sometimes quickly—proved wrong, let me assemble evidence that could be seen as supportive.

First of all, there's a lingering rumor that Walt Disney was frozen immediately after death. The persistence of the rumor relies on the idea that if Walt believed in it, maybe it's true (although don't forget he also came up with Fantasyland). Actually, he was cremated. Nevertheless, this illustrates that cryonics has been in the public mind for decades, and that's an important component in belief.

Evidence trumps belief, however, and while there is some suggestive evidence, it's anything but concrete. So, for instance, when Arctic ground squirrels hibernate, their body temperature drops dramatically, sometimes to below the freezing point of water; parts of their brains shut down, and they remain in that near-death state for months. But when spring comes, they are back to life in every way. Even more dramatic are wood frogs, which can practically freeze solid in the winter. They too are fine come spring. However, dogs and cats are very different from ground squirrels and frogs, and those animals are only in the supercooled state for months, not centuries.

Cats and dogs are closer to humans, and there are a few human examples too. In 1986, a two-and-a-half-year-old girl fell into a creek and wasn't rescued for more than an hour. She had been submerged all that time, and her temperature dropped to 22 degrees Celsius (72 degrees Fahrenheit), fifteen degrees below normal body temperature. She wasn't breathing, her heart wasn't beating, yet after a series of intense medical interventions, she recovered, and other than a slight tremor in her hands, she is pretty much normal today. But 22 degrees Celsius isn't –196 degrees Celsius (–321 degrees Fahrenheit), the temperature of liquid nitrogen at which cryonically preserved bodies would be kept.

If you opted for this, the body of your deceased pet would (hopefully immediately) be cooled to be about –120 degrees Celsius (–184 degrees Fahrenheit). At that point, body liquids are between liquid and solid in consistency. This ensures that they won't be damaged as cooling continues to –196 degrees (–202 degrees Fahrenheit).

A pet's entire body could be preserved this way. But we don't know yet if, upon revival, it would be completely intact. The simpler the organ, the likelier it could survive. Already scientists have been able to freeze mouse ovaries to –196 degrees, thaw them, and transplant them into female mice and produce live, normal babies. Even slices of rat brain have been cooled to –130 degrees and, when warmed, seem to be functional, but after all, they're only slices.

The uncertainty of recovery is one thing, but the time span is another. No one has any idea when the technology of resuscitating a frozen body might be developed, if ever. The longer the wait, the more uncertain the entire process becomes.

Alcor, the US-based cryonics company, wants anywhere from $30,000 to $130,000 to freeze your pet's entire body, depending on the sophistication of the technology used. The pet's brain can be frozen on its own, in the hope that when the technology is ready, the brain can be thawed and a new body created, either an organic body from the pet's DNA or a completely synthetic version, or maybe the pet's brain will exist virtually, with no body at all. How the latter would be a satisfying companion is beyond me. Also, the pet owner must be registered with Alcor as well ($$$), so it's a team project.

And an elaborate one. A Utah-based startup called Cryopets, funded by the Silicon Valley billionaire Peter Thiel (who also wants to be frozen when he dies), claims to have a wait list of five hundred animals. The founder, Kai Micah Mills, in an interview with the *Daily Mail*, outlined the elements of the process.[5]

Using the example of an old dog dying of cancer, Mills lists the necessary procedures in order: euthanasia, immediate cooling, immersion in an ice bath for continued cooling, implanting a device to prevent blood clotting as it's replaced by cryoprotectant (similar to antifreeze), continued cooling, and finally placement in a cooldown box to lower the temperature to its final setting.

It's important to note that no one has actually tried to freeze a dog solid and revive it (or at least no one has admitted doing it). There were experiments in the late twentieth century lowering the temperatures of dogs dramatically and bringing them back, but the lowest temperatures were nowhere near what's proposed for cryonic preservation.

The typical game plan here must be to freeze the pet when it dies; then freeze the pet owner; then, centuries from now, thaw them both out at the same time. You both come back to life in a totally different world. What if pets are banned in that world? What if humans of the time see dogs, or cats, as edible items? What if the pet is revived and you're not? Then the medical-miracle animal might end up in a shelter, if indeed there are shelters in that future world. Or euthanized. That would be ironic.

Honestly, I think taxidermy would be preferable.

– CHAPTER 28 –

Could There Be Robot Pets?

If your first reaction to the question posed by the title of this chapter—"Could There Be Robot Pets?"—is "This is ridiculous," I'd remind you that there are already robot pets. They're basic, yes, certainly not convincingly "pet-like." But I'm building a case not for them but for their future relatives. The following connect-the-dots is speculative, with many twists and turns, but I think it's plausible. Whether you'll agree might depend on exactly what kind of relationship you already have, or have had, with a pet.

There are millions of people who count their pets as "members of the family" (50 percent even consider the pet to have equal rank with the humans); there are also millions whose relationship with their dog or cat isn't quite that intimate. Important to remember, as this story unfolds, that some pet owners might be satisfied with something that closely *resembles* a living animal but isn't quite.

As I've said, today's robot pets are not close to pets at all, but are small, mechanized, sometimes furry objects that emit sounds and can provide some degree of comfort and companionship—in a pet-like way—to people who need it. They have been shown to be especially effective in providing company, or at least a presence, for people living with dementia.

Apparently, the first robotic pet was Sparko, designed by Westinghouse to accompany their humanoid robot Elektro at the 1939 World's

Could There Be Robot Pets?

Fair. Sparko was never actually put into production due to a lack of public interest. The idea of robotic pets didn't take hold until the late 1990s, when Hasbro's Furby and Sony's Aibo hit the market.

Some of the first studies of children playing with Aibo revealed they had mixed feelings about the little robot. Attitudes toward Aibo were age-related: Preschoolers expressed reservations about Aibo as a living pet but at the same time *acted* as if it were, while seven-to-fifteen-year-olds clearly rated a living dog higher but still found room for acknowledging that Aibo was doglike in some ways. Even a third group, adults, behaved from time to time as if Aibo were alive, but sometimes as simply a piece of technology.[1] Another research group, having found similar results, put it this way: "While in one sense the children knew that Aibo was an artifact, that knowledge did not stop them from conceiving of and treating Aibo socially in some ways *as if* it were a real dog."[2]

It's important to remember that this is the furthest thing from a living dog—it's a gray metal object with flashing lights that makes musical sounds.

From these early studies with children, much more has been done to investigate the reactions of adults. One of the most consistent observations has been that robot pets, at the very least, capture people's attention. But it doesn't stop there: They also have a positive impact. Being with a robot pet makes people, especially seniors, happier and less anxious, results positive enough to give some impetus to the conversation about how widespread robotic pets might become. Could they end up as staff live-in pets at long-term care centers? One of the most popular robots in this research is the seal called Paro.

Why a seal? A direct comparison of experimental cat and seal robots led to complaints from cat owners that the robot feline wasn't realistic enough. But people weren't familiar with what seals were supposed to feel like, so they couldn't critique the seal robot on the basis of realism. And so was born Paro, the white harp seal.

Thinking Long-Term

Paro is a step beyond Aibo: It is tech-heavy, with dual 32-bit processors, three microphones, and tactile sensors (whiskers). It moves its legs and arms, opens and returns your gaze. It also cuddles and makes baby-seal-like sounds. Is this the best demonstration ever of biophilia—or the worst? It recognizes its own name and responds to encouragement and (*gasp*) punishment. Stroke it, and it will repeat what it just did; hit it, and it will try not to.

It works. People become strongly attached to pets like Paro, although at the same time likely displaying those contradictory thoughts and actions I mentioned earlier. A study in England revealed that Paro succeeds on many levels, including improving social interactions and encouraging attachment and the recall of pleasant memories, all in addition to the physical interaction.[3] Even those living with dementia pay unusual attention to Paro. The plump 2.7-kilogram (6-pound) seal elicits comments of love and devotion.

Kathy Craig is a therapist at the Livermore, California, VA hospital. "We'll bring out the Paro robot and set it down and they'll start talking to the Paro, they'll talk to other people, it'll brighten their mood. And if they're maybe at risk of wandering and getting lost, instead of that happening, they might sit down with Paro for a while and spend some time with it."[4]

Let's imagine that "companion pets" like Paro settle into that niche, and other, more elaborate and entertaining robopets spread out to the rest of society. Would a proliferation of Paros threaten the bond humans already have with their (living) pets? It's a lot easier taking care of a robopet, and they're just as faithful.

Professor Lisa Carver of Queen's University in Kingston has a practical solution for the "Let's have robots"/"No robots" standoff: Yes, robot dogs, but only as *service* dog robots. She calls them "Serve-U-Bots," robotic pets loaded with sensing technology that could replace living dogs that do the same job now, but much more expensively. It

Could There Be Robot Pets?

costs Canadian Guide Dogs for the Blind almost $75,000 to train each of the twenty-three dogs they graduate each year.

Of course such robodogs could eventually be programmed to take over many of the tasks assigned to living dogs today, like crisis response, search and rescue, airport-runway wildlife control, even sled pulling, although it will be quite a while before robots can do olfaction-dependent tasks like truffle hunting and disease detection as well as a dog.[5]

Exactly what incentives would have to come together to make this happen I don't know, and maybe it never will, but it is just as easy to imagine robopets in the future as it is to dismiss them. Because there's a bond.

Professor Carver doesn't see her Serve-U-Bots as companion animals at all, but strictly for service. And yet the human desire to connect, biophilia married to anthropocentrism, can expose emotional connections in the strangest of circumstances.

Kate Darling, in her book *The New Breed: What Our History with Animals Reveals About Our Future with Robots*, tells a story about an emotional attachment. The US Army had been testing a multi-legged robot built to destroy land mines. Whenever it located an active mine, the mine would detonate and the robot would lose a leg. However, it would stagger on. "But the army colonel who was overseeing the testing exercise ended up calling it off. According to the reporter, 'The colonel could just not stand the pathos of watching the burned, scarred and crippled machine drag itself forward on its last leg. This test, he charged, was inhumane.'"

This isn't the only story of extreme attachment to a robot in the military that Darling tells. In another, a letter of condolence was sent from an officer in the navy to the manufacturer of a ruined robot, praising its "sacrifice."[6]

This range of stories, from long-term care residents cuddling their robot pets to military men treating twisted metal with the respect that an animal (or a human?) would get, is pretty convincing that the barrier to engagement between a robot and a human is lower than you might have thought. However, it is probably worth retaining some sort of barrier. Maybe even an ethical one.

When it comes to robot caregivers, humanoid or pet, Robert Sparrow, a philosopher at Monash University in Australia, has some issues. He objects to the fact that by giving such robot pets to people, we are either deceiving them or asking them to participate in their own self-deception—either is immoral. And while he admits robot pets are easier to care for than the real thing, he wonders if they provide the same kind of intimacy and comfort the real ones do. He extends this argument to foresee that the introduction of humanoid robots in the care of the elderly, a practice that many have argued would be beneficial, would in a similar way reduce the respect and recognition central to such care.

Any time you encounter the issues that arise when technology meets humanity, Sherry Turkle has something to say. The MIT prof and author has thought a lot about robotic companions. She had a blunt assessment of a video extolling the virtues of Paro. In the video, the seal was with a woman named Miriam. "To say that Miriam was having a conversation with Paro, as these people do, is to forget what it is to have a conversation," said Turkle. And "the very fact that we now design and manufacture robot companions for the elderly marks a turning point. We ask technology to perform what used to be 'love's labor': taking care of each other."[7]

Turkle argues that since there is no empathy directed from machine to human, such relationships undermine the fabric of person-to-person relationships. In her view, there's an emptiness: "the illusion of intimacy without the demands of friendship."[8]

All this is objectively true, but if, for instance, the only companionship available to someone is artificial, and it relieves the symptoms

Could There Be Robot Pets?

of loneliness, is that wrong? Acknowledging the stress of being a caregiver, there might be that day when, at the end of a long visit, there's a temptation to turn the visit over to the "pet." As the pets get more sophisticated—and therefore more convincing—it might be more and more challenging to ensure that there's a human somewhere in all this.

- PART VIII -

The Future

– CHAPTER 29 –

Talk to the Animals

In the 1967 movie *Doctor Dolittle*, the doctor himself knows five hundred languages spoken by land creatures and birds and is learning additional marine languages to prepare for his search for the Great Pink Sea Snail. Dr. Dolittle was a cinematic fantasy, but in the years since, things have changed. The combination of a much deeper understanding of the signals animals send to each other, a greater appreciation of their brain power and behavior, revelations of the communication between humans and their pets, and especially the incredibly rapid growth of AI's large language models have set the stage, at least in some scientists' minds, for a revolution in communicating with our pets.

"Talking to the animals" was on the minds of Yossi Yovel and Oded Rechavi when they came up with the Doctor Doolittle challenge.[1]

It is now called the Coller-Dolittle Prize for Two-Way Interspecies Communication, as money from the Jeremy Coller Foundation has made it possible to create prizes for winning the challenge. The grand prize is either a $10 million equity investment or a $500,000 cash prize, but there will also be annual prizes to push the advancement of new ideas and approaches.

Yovel and Rechavi wanted to create "CatGPT" to allow cat owners to converse with their cats, and they set out the goals they think would have to be achieved to be able to brag that communication had effectively been established. There's a similarity to the famous Turing test (can you tell whether it's a human or a computer you're conversing with?) in that the goal of the challenge is to communicate with animals without the animal knowing it's "talking" to a human.

The challenge, in three parts, is to develop a machine that

1. communicates with an animal using *its own endogenous signals*;

2. does so in various behavioral contexts; and

3. works such that the animal *produces a measurable response* to it as if it were communicating with a conspecific and not a machine.

Each step is an enormous challenge. The scientists specified the animal's "own endogenous signals," meaning that a signal native to the animal must be used. So if the animal involved is a dog, a human "command" could not be a spoken word, but rather a bark or a growl. Saying "sit" would not qualify; it would have to be "sit" translated into canine, and I doubt there actually is a way of doing that. Has any dog ever ordered another dog to sit?

The "various behavioral contexts" demand requires that whatever is communicated to the dog must be done in a variety of situations, presumably including settings like alone in the house, walking outside, even in a dog park. And finally, the dog must respond in a way that demonstrates it has understood the communication. The authors put it this way: "It should, for example, escape when we broadcast an alarm signal and approach or vocalize when we generate a contact signal."

And while this isn't specified, surely there would be a premium put on a long, extended exchange between dog and human before anyone would really be impressed.

The authors point out—encouragingly—that AI has already solved the puzzle of generating animal sounds from scratch (they've done that themselves with the vocalizations of fruit bats). But there is still a steep hill to climb. One of the crucial things is the way a pet would respond: "Notably, to pass the Doctor Dolittle challenge the animal should respond without learning, even when exposed to our signal for the first time." Without learning, because the language used is one in which they are already adept.

The Doctor Dolittle challenge is out there. Are the authors of it optimistic? Yes . . . and no.

The giant leap in AI has dramatically enhanced our ability to encode data and perform predictions on unseen data or even generate new data. These advances can also be applied to animal communication. They will likely yield a better understanding of animal communication, but they will not necessarily enable communication with animals.

That distinction is important. We are, having lived with dogs and cats for millennia, somewhat familiar with their efforts to communicate with us and even within their own species. We're pretty sure we can interpret howls, barks, growls, and hisses. At the same time, scientists are beginning to be able to interpret the signals wild animals exchange in the absence of humans.

If we made significant progress, it's hard to imagine what might result. Would animals (including pets) always tell the truth? Would they explain fundamental truths about themselves that had never occurred to us? Would they be shy to admit things they suspect you wouldn't like to hear? The possibilities are endless. But there is still a very long way to go.

– CHAPTER 30 –

The Wild World of Communication

There is widespread interest in the idea of talking to your pet—always has been—but if we're honest, it would be more important to gain insights into the communications of wildlife. We live in an era of unprecedented pressure on the world's species, living as we are in the sixth mass extinction.[1] If we did know more about how endangered species communicate, we might be able to make better environmental, ecological, and biodiversity decisions. And we are beginning to make progress.

The first step could be simply knowing more about the habits of some wild species. For instance, in Namibia, AI is helping to identify individual brown hyenas from photographs of their footprints, allowing scientists to track and identify these elusive animals without actually having to be on the scene.[2]

The next step is to watch and listen. Over the last few decades, there's been an onslaught of observations and data pointing to elaborate communications among wild species. The variety is incredible: Male elephants send the LGR ("Let's Go" Rumble, as in "This is the rumble that says 'Let's go'") message to others near the water hole at frequencies so low they are inaudible to humans, while female elephants watch to see if others are looking at them before choosing appropriate combinations of gestures, like ear waving, and sound to communicate.

It's been known for decades that sperm whales use clicks and other sounds to communicate, and recently researchers analyzed the clusters of repeated clicks, called codas, and found that what matters is more than just the length of a click or the sequence of clicks, but also some audio features that depend on the "conversational" context, and some that don't. Ornamentation, rhythm, and tempo are some of the coda features they identified and named. It sounds like a grammar and a vocabulary. (In fact, additional research suggests language-like structures to their sounds.) Of course sperm whales, the animal with the biggest brain in the world, should be expected to be expert communicators, given that they spend much of their lives submerged in the complete darkness of the deep ocean. On the other hand, there's no reason to expect their communication is anything like ours, given how different they are. These are two very different examples, but they share features that convincingly argue that they're communicating, while still leaving us puzzled as to exactly what. Certainly, at this point, no one can claim it's exactly "language" in the human sense.

There are potentials applications for this knowledge: Crows that escaped extinction in captivity are now being returned to their native Maui, but there are fears they might have lost their natural communications calls in the interim—knowing the meaning of those calls might open up the possibility of "teaching" them to the crows. Or understanding the complete repertoire of beluga whale sounds in the Saint Lawrence River might enable informing passing ships when the whales might be surfacing. But there are potential downsides too: Easier tracking and better insights into how—and what—animals are thinking can open doors to abuse. Knowing the emotions and intentions of animals might allow us to gain even more control over their behavior. Think training dolphins to clear mines from harbors. Or, once AI is familiar with the range of chicken sounds and their

The Future

meanings, anxiety-reducing chicken sounds could be broadcast in the poultry house. More eggs! Fatter chickens! Maybe this is a wild and unlikely scenario, but imagine elephant poachers interpreting the sounds of the herd to anticipate where they're going. Or communicating with them to encourage them to approach.

There would be a need for caution even at the earliest stage of technology development. Imagine a team is experimenting with a collage of newly synthesized sounds like those used by whales. They are sure they understand what each sounds means, but they have to test them in the wild, their natural setting. After broadcasting them, the scientists realize that their sounds, rather than being seamlessly integrated into the pod's behavior, have disrupted it.

There's time to consider these risks because we are still far from a comprehensive understanding of the communications of wild animals. In fact, a large part of what we've learned about their capabilities has come from individuals in captivity.

A trio of captive animals are generally considered to have been the all-stars of communication: Koko the gorilla, Kanzi the bonobo, and Alex the parrot.* Each appeared to have a sophisticated capacity for understanding spoken (or sign) language, together with the ability to generate it. An important qualification is that all three were good at human language, and little was learned from any of them about how each species communicated with its own. Nonetheless, their capacity for turning thoughts into symbols is what made them most impressive and at the very least suggested it might be possible to communicate with them according to the guidelines of the Dolittle Prize.

We met Alex the parrot in chapter 11. Just to recap: He knew about a hundred words to describe a range of colors and a variety of objects. Apparently, when he first saw himself in a mirror, he asked: "Color?"

---

* Sadly, all are now dead.

The Wild World of Communication

He mastered the answer "gray" after being reminded half a dozen times. And Alex's owner and trainer, Dr. Irene Pepperberg, once told me on *Quirks & Quarks* that Alex, if bored, would even deliberately answer questions incorrectly—mischievously.

Koko the gorilla had the best PR. She didn't just know two thousand spoken English words, or have the ability to use American Sign Language (one thousand signs); she was tight with some big names. Jane Goodall remembered her as a jokester, and Koko was especially close to the comedian Robin Williams. She was visibly upset when told he had died.

Koko is obviously a stellar example of an animal that can understand language. Or is she? Penny Patterson's research work with Koko was always under a bit of a shadow, even as the research was underway, because very little peer-reviewed evidence ever appeared. Lots of fanfare but short on the science—that was the argument. But even if everything reported for Koko actually happened as described, that wouldn't mean that Koko was capable of "language." Learning a sign for an object versus being able to express a continuous flow of thoughts using signs—they can hardly be compared as the gap between them is so huge. That isn't to say Koko wasn't remarkable—knowing there's a gorilla that can communicate with words or signs takes us much further than we were. But there's still a long way to go.*

Kanzi the bonobo could apparently understand complex commands, like when asked to "get the carrot in the microwave," he would

---

* I'm not surprised by any evidence of primate braininess. Once, on a trip to Japan for Discovery Channel, I was pitted against a chimpanzee named Ai in a memory game. Nine digits appeared in random squares on a screen, and you had to touch them all in sequence, 1 through 9. But as soon as you touched the 1, the others disappeared, leaving only blank squares, and you had to touch them in the correct order nonetheless. She killed me—it wasn't even close. Here's an example of a different chimp attempting the same puzzle: "Superior Chimpanzee Memory," Cell Press, December 10, 2018, YouTube video, https://www.youtube.com/watch?v=PNrWUS13th8.

ignore a carrot directly in front of him and go straight to the microwave. He also apparently had learned a huge number of words. When a word was spoken, Kanzi would point to a symbol that meant that word. Two great stories from Kanzi's life: When he was eight, he and a two-year-old girl were tested for their abilities to understand spoken commands. Kanzi outscored the girl.

Several years later, Kanzi was taught how to make stone tools from scratch in the old Ice Age way. The development of toolmaking in human evolutionary history is thought to have been a crucial advance, not only because well-made tools made life easier, but because the techniques changed the wiring of the brain, perhaps even connecting to the invention of language. Kanzi was pretty good at it and even invented new techniques. It's hard to impress people, though: The scientific jury concluded that the authentic Stone Age tools from back in the day were better. Kanzi's experimental life was much more thoroughly documented scientifically than Koko's, but that wouldn't resolve the questions. Most crucial of which is: Is all this really language?

Each of these three was an astounding example of an animal understanding words in English, and each was accompanied by claims that its thought was at the level of a human infant or higher. But, as you'd expect, these claims were also controversial. From everything we know so far, human language is vastly more complicated than the efforts at language of these famed three. Critiques ranged from too much anthropomorphism in interpreting what was said or signed, to claims that answers were tipped off by the kinds of question asked, to inadequate experimental design that heightened the risk of false positives. In particular, in Koko's case, critics focused on the shortage of peer-reviewed evidence.

But what these animals were able to do shouldn't be downplayed either: They opened scientists' eyes to what might be possible.

It's clear then that animals communicate with each other in subtle and complicated ways; it's also true that some special, brainy animals like parrots and primates can handle some aspects of human language, without having nearly the facility with it that we do. (But that can be said of cats and dogs too!)

There's no doubt that parrots, chimps, and sperm whales are unusually smart. And all you border collie owners, just stay calm when I suggest they are smarter than dogs or cats. But how widespread is "intelligence" (a very slippery word) when it comes to other, possibly underappreciated animals? I'd just offer the example of Aesop's fable "The Crow and the Pitcher":

> In a spell of dry weather, when the Birds could find very little to drink, a thirsty Crow found a pitcher with a little water in it. But the pitcher was high and had a narrow neck, and no matter how he tried, the Crow could not reach the water. The poor thing felt as if he must die of thirst.
>
> Then an idea came to him. Picking up some small pebbles, he dropped them into the pitcher one by one. With each pebble the water rose a little higher until at last it was near enough so he could drink.[3]

Smart crow, but a fable, right? No, real crows seem to be able to figure this one out, as well as others involving a series of decisions, like using a short stick to get at a longer one, to get an even longer one, to finally pull a treat out of a cage.[4] This is only one of many demonstrations that crows can analyze a problem and create novel solutions. It wasn't that long ago that scientists were dubious about any claims of bird intelligence outside of parrots.

What about prairie dogs? They're not known to be brainy. Con Slobodchikoff of Northern Arizona University is a name in this field; he has characterized the surprisingly complex vocabulary of prairie dogs, showing that the variety of calls coded not just for the identity of a threatening predator but also for its size, shape, and color. Prairie dogs will also tailor their escape behavior to the specific predator that they have identified, even without seeing or smelling it. (Such linguistic prowess is ironic when some of the most popular YouTube videos mock the incessant utterances of prairie dogs.)[5] Slobodchikoff is not content to stop with prairie dogs, however: He has said that he wants to create a "Dictionary of Barks," which, with the help of some dedicated AI, could translate a human command into a bark that the dog would immediately understand. Or vice versa: a bark becoming a shouted word.[6]

Slobodchikoff deserves credit for his revelations about prairie dogs (his most recent paper shows that they have dialects!), but even he admits that the Dictionary of Barks is a long way off.

I think there's enough evidence here to say there's some chance (remote?) we might be able to understand the thinking behind the communication efforts of other animals. There are obvious reasons why we might make better headway with pets. We understand them best. And if millions of pet owners could be recruited to act as citizen scientists, an unprecedented experimental campaign could be launched. Maybe, just maybe, we'd see the day when the Dictionary of Barks is published.

– CHAPTER 31 –

You Talkin' to Me?

We know our cats and dogs best. We already communicate with them, at least in a very basic way. A reminder, though: We're communicating with a handful of gestures, but mostly words. But we're talking *to* them, not conversing *with* them. For that, we'd need to be able to communicate the way they do with each other. We're not there yet, but we're moving in that direction.

For example, there is some scientific literature (and thousands of TikTok videos) suggesting that augmentative interspecies communication (AIC) devices illustrate that dogs can understand human speech and act on it. A dog owner will immediately say, "Of course they can: My dog knows immediately what's up when I say 'dinner.'" Yes, your dog does, because you're standing there with a full food bowl in the exact place where you always feed him/her. You're providing multiple cues (not unlike the Clever Hans story from chapter 10).

For this to be a convincing demonstration of a dog's language comprehension, the word "dinner" has to be the one and only cue, divorced from anything else the human does or says. A recent experiment demonstrated that there is promise in using AICs, colloquially called "soundboards," where a variety of recorded words, in the owner's voice, like "out" or "outside," "play" or "toy," or "food" or "eat," will sound when a button is pressed and prompt a trained dog to act accordingly: go to the door or the toy bin or the dish.[1]

But eliminating extraneous influences is hard. Where does the owner stand when the word is heard? Out of sight? What is the owner doing at the time? Does it have to be the owner's voice? If so, then the dog cannot be said to be understanding the word itself. In this particular experiment, dogs behaved differently in response to the words "play" and "outdoor." Upon hearing "play," a typical response is this: Dog approaches toy, picks up toy, plays with toy. But when the cue is "outdoor," in the somewhat indirect words of the researchers, "distance between dog and door decreases." But when the word was "food"? Nothing. No demonstrable behavior, like moving toward the food dish or pestering the owner or barking. Nothing. Unbelievable, but the researchers again: "We found no conclusive evidence to suggest that dogs exhibited food-directed behaviors." Imagine—if you can—dogs not responding to the word "food," doing nothing to show that they might understand the word. The research team did comment that perhaps the dogs had been fed shortly before the experiments, or that perhaps some dogs weren't expecting food outside of their normal schedule. Really? Those do have the ring of attempts to explain away unexpected data. I mean, I do know one or two dogs that could probably abstain, but mine isn't one of them.

One of the cool things they did show was that it didn't matter whether the owner or a stranger spoke the words, or the owner or a *recording* of the owner spoke the word. It seemed like the dogs understood the meaning of the words. It was also clear that pet owners on their own came up with results comparable to that of the scientists, suggesting that a massive citizen science research program I suggested earlier might be feasible. And while there *is* a flood of TikTok videos portraying smart dogs having "conversations" using a soundboard, so far there's only one dog, a mongrel named Sofia, who has been shown to make requests like "pet me" or "give me water" in scientifically airtight situations—no hints being dropped.[2]

But as provocative as this work is, we're still a long way from being able to communicate with dogs (or any other animal) on the animal's own terms. The common element here is human language—it's not what animals use. But momentum to dig deeper is definitely building.

One impressive kind of experimentation involves imaging dog's brains with functional MRI. Impressive mostly because the dog in the MRI apparatus has to remain perfectly still; otherwise fuzzy images and a spoiled experiment are the result. But there actually are dogs that will remain still during an MRI study for thirty minutes! This has made possible the following insights:

When dogs are presented with a mix of visuals of objects (dog, car, human, cat) or actions (sniffing, playing, or eating), they pay much more attention to actions. This is an important difference from humans, who are very object-oriented.

My favorite experimental design featured a dog in an MRI machine being tested with three different objects, paired with three different rewards. A miniature car, held about two-thirds of a meter (two feet) in front of the dog's nose, was followed by three seconds of praise; a toy horse followed by a chunk of wiener; and finally a brush with absolutely nothing as a reward.

The MRI revealed areas of the brain that were activated by each object, and thirteen of fifteen test dogs showed about the same excitement with praise as they did with a hot dog. To say, as the authors do, "that the majority of our participants found social interaction at least as rewarding as food" took my breath away. Another example of dogs not being totally obsessed with food? Dog owners, do you believe that?[3]

Another brain-focused MRI revealed a left- versus right-hemisphere bias: The left hemisphere of a dog's brain is responsible for most of the processing (and thus understanding) of words that are meaningful to the dog ("treat"), but the right hemisphere predominates when understanding the significance of intonation (and thus emotion). But neither would be useful on its own, and sure enough, the two canine brain

The Future

hemispheres conspire to give their owner the full picture: The reward areas of the dog's brain, which are well known, were primarily activated only when meaningful words were uttered with a positive tone.[4]

It's fair to say that you probably suspected that a dog would get most excited when a familiar word is uttered with a positive tone, so in that sense, we might not have learned anything radically new. But there's now a familiar path in the dog's brain that activates when you'd expect it to, so at least that's reassuring.

But there's still an enormous gap between what's been done and what needs to be done. And it's not even really clear what *does* need to be done. Are we really trying to understand an animal's *thoughts*? We can't even be sure they have thoughts that we could recognize as such.

A personal example: My dog Robbie eagerly chases a ball, but rarely catches it first lunge. Once he has caught up to the ball, instead of grabbing it immediately, he stands over it for at least four or five seconds—or more—then pounces. I want to know what's going on in his mind while he's waiting. Is he afraid of something? Does he worry the ball has acquired some amazing escape skill since he last mouthed it? Why wait?

It might be that even with the most powerful and insightful AI, Robbie's thoughts exist in a format I could never understand. Jason Beck of York University in Toronto puts it this way: "Trying to characterize animal thought is like trying to describe the Mona Lisa. Approximations are possible, but precision is not."[5]

Temple Grandin is famous for being able to look at difficult situations in animal husbandry through the animals' eyes, an ability she attributes to her autism. In fact, she has argued that animals are like autistic savants, those rare humans who have one, maybe two, incredible abilities, whether it is to remember every note of a musical piece in one hearing, calculate the day of the week you were born (in a second), or draw scenes in detail from a glance. These extraordinary individuals perform their one skill at superhuman levels, but often can barely manage the rest of daily life. They are both extraordinary and less than ordinary.

Grandin argues that autistic savants and animals are alike in that both have special talents. Autistic savants demonstrate skills that are unfathomable. She believes there is common ground there, that other animals have extraordinary skills too, of which we are unaware.* As she puts it: "Animal genius is invisible to the naked eye."[6]

Maybe we're just a little naive to think that throwing some technology at animals' brains will reveal what they're thinking.

---

* Dogs and cats pay attention to television inconsistently, but recent reports suggest that dogs are unsettled by the hit series *Severance*, and both animals seem preoccupied by the Oscar-winning animated movie *Flow*. The suspicion is that high-pitched dings and card beeps in *Severance* and the representations of real animals in *Flow* attract their attention—but we don't really know for sure.

– CHAPTER 32 –

A Cautionary Tail

> She now stands upright, with slightly arched back, tail perpendicularly raised, and ears erected; and she rubs her cheeks and flanks against her master or mistress.
>
> —Charles Darwin, on the movements of an affectionate cat, in *The Expression of the Emotions in Man and Animals*

The soundboard research I described in the last chapter has delivered encouraging insights into the world of dog communication. There's also no doubt that we understand our dogs and cats more than any other animals. But even the soundboard research relies heavily on the symbols and noises we call language, and it's just possible that we might be taking a reckless step in believing that's what we should do with animals, even our pets. A Dictionary of Barks, or Meows, might be overlooking some crucial facets of cat and dog communication. Like body language. The need to take the broadest view possible can easily be illustrated just by looking at our cats' and dogs' tails.

It's worth recalling how we domesticated dogs and cats, and the impact that process had on how we communicate with them today. Cats insinuated themselves into our lives with their useful independence. Millennia ago, people began storing grain, grain attracted rodents, rodents attracted cats, and a beautiful relationship was born. It's commonly said that cats are "half domesticated," and the widespread

A Cautionary Tail

existence of feral cat colonies around the world attests to that. Countering that is the obvious fact that millions of cats live very domestic lives, cuddling with their humans (and occasionally bringing a dead bird home as a treat). The wildcat, from which house cats descend, is solitary; the fact that house cats have become social with other cats and humans is a tribute to their social flexibility.

Dogs got off to a very different start. They've been around us much longer than cats have, and spent that time accommodating to nomadic hunter-gathering humans, establishing firm communication and emotional bonds with them. They didn't have to develop that familiarity with other dogs, as cats did with cats, because they were pack animals to begin with.

They're now very different from wolves, highly socialized, yet in the transition from wolf to dog, something was lost: the lupine face. Dogs' muzzles are shorter, teeth have shrunk, ears have flopped, lips have protruded, and the effect of these cumulative changes over centuries suggests that dogs might not be able to display emotions facially as vividly as wolves do. A recent experiment comparing the two confirmed this. It was much easier to derive emotional states in wolves (fear, anger, curiosity) from expressions (featuring eyes, ears, teeth, and lips) than it was in dogs. The authors did note that in an effort to counter this shortcoming, dogs appear to rely more on vocal communication.[1] However, a recent experiment showed that humans—or at least the brains of selected humans—process dog facial expressions the same way they process equivalent human expressions, with more empathetic humans making more accurate judgments of both species' emotions.[2]

Much of the scientific focus on the wolf-to-dog transition has been on the polishing of the dog's ability to respond to humans and vice versa. Yet here we see that breeding targeted facial features has diminished the clarity of the dog's emotional expressions. Cats also rely on facial expressions with other cats—and humans—although there are differences between the two.

The Future

If you're going to take the idea seriously that we might one day communicate much more efficiently and deeply with dogs and cats, then it'd be worth knowing as much as possible about what goes on in their heads, before you try to enhance communication. Not just knowing that sound X means Y, or this gesture triggers that behavior, but the intent and meaning behind different communications.

Also, "animal talk" could be assumed to be vocalizing. The scientific classification of cat vocalizations includes caterwauling, chattering, chirps, hisses, howls, moans, murmurs, and many more. But sounds are only part of cats' and dogs' communication package. Dogs are famous for their olfactory-detection abilities, and in both species, scent can signal territory, good health, and likely many attributes we haven't yet identified. Body language is expressive too and a little easier for humans to deduce. Even when you compare the same parts of the body, the differences between cats and dogs are stark.

The Charles Darwin quote at the beginning of this chapter refers to the familiar straight-up tail of the cat. Most cat owners know that if their cat enters the room with tail held straight up, or even with a slight kink near the top, that's a good thing. The cat is signaling contentment, either with its own state or the scene in front of it. Or both. This is a direct contrast to tail down or between the legs, a defensive signal. Unsettled cats sometimes lash their tail back and forth; usually that's a sign of fear.

However, whether a cat deploys the tail-up signal because it wants to approach and rub depends on the cat. All domestic cats do it, but feral cats are not known to use the straight-up tail except to facilitate spraying (scent is a strong signal). This suggests that transitioning the tail-up pose from spraying only to communication and spraying likely happened after cats were domesticated.*

---

* It has been suggested that the tail-up signal might have originated in ancient Egypt, where cats were being treated royally.

A Cautionary Tail

But just to complicate things, the tail isn't on its own. At the same time, the cat's ears are adding to and modifying the signal (as Darwin noted); straight-up ears anticipate positive encounters, say with another cat, while the more flattened the ears (possibly to protect them in the case of a fight), the more negative the signal. Straight-up ears and a tail like a flagpole together are friendly gestures, at least cat-on-cat, at least with domestic cats. Feral cats seem to depend on the ears instead. In one study, the outcomes of one-on-one interactions, good or bad, were signaled by the ears, not the tail. The background to all this is that feral cats are solitary most of the time, and largely nocturnal, so visual signals likely don't play as significant a role as scent does.[3]

Cat-on-human is a different social situation. An upright tail is obvious to the most casual cat owner, where the angle of the ears might not be. The tail seems to have taken over as the main signal when humans are the target. It might seem straightforward for AI-related tools to sort through all the varied visual dynamics of body posture, tail angle, and ear position to gain a better understanding of what a cat is thinking. But what if some of these postures are bluffing—could AI detect that? Do cats bluff?

The upright tail is the cat's social display of choice, at least with humans. The dog's is the wag. And no, the tail does not wag the dog—I have evidence.[4] But if, in your mind's eye, you're seeing the gentle swish-swish of a dog's tail back and forth, left to right, as a simple "happy to see you," you're missing some cool variations on that theme.

The "Tail Wags the Dog" report I alluded to above has not yet been peer-reviewed (as of time of writing), so you could be cautious about the findings, but it makes the case that dogs hardly use their tails *at all* for any biomechanical advantage, whether that's balance, energy saving in turns, body posture, or acceleration. For animals like cheetahs, the tail is an essential part of the chase package—but apparently not for dogs. Faced with needing a reason for dogs having tails at all, the scientists who did this study concluded that dogs use their tails either to whisk away annoying insects or for communication. I'd put most of my chips on communication.

An animal like a dog must have several channels of communication, with flexibility built into each. My dog Robbie doesn't have just one bark. He will always bark if anyone comes to the door. If he knows the visitor, the bark is pretty loud, but not frantic, about an 8 (although he leaps into the air repeatedly). Frantic is when it's someone he doesn't know—that's an 11.

The same is true of tail wagging. There's the obvious side-to-side movement, but there's also the angle: upward, level, or drooping. Each of these positions has a meaning and must be taken into account if you're to read the demeanor of the dog attached to the tail.

There is general agreement that a dog with its tail held high is on the confident/aggressive part of the spectrum. To some extent, the closer to vertical, the more threatening the situation. Level is neutral, while a downward tail can be a submissive/fearful. But there are variations in exact positioning and timing, complicated by the presence of other body signals and environments. Now add the wag.

Imagine you're standing behind a dog facing the same direction he is. His tail is level, wagging back and forth, right to left, with windshield wiper regularity. Straightforward and simple, but it is easy to change that sweep of the tail simply by adding something to the dog's environment, like an *unfamiliar* person entering the room.

The wagging would likely intensify, but more interesting is that the tail would likely move farther to the left. If it had been moving through the air like a clock hand—from, say, eight o'clock to four o'clock and back again—when the human enters the scene, the tail would flip to something more like ten to six. It's an instant expression of uncertainty.

Wagging left, when the tail spends more time on the left side of the dog's body, is possibly unfriendly, but at a minimum cautious.

A Cautionary Tail

Wagging evenly left and right is neutral. Wagging to the right (the dog's right) is positive emotion.

There are straightforward ways of demonstrating this. In one experimental setup, dogs viewed four different videos: the dog's owner, an unfamiliar person, a cat, and an aggressive, dominant dog (dominance established by standard behavioral tests). Their tails revealed their thoughts: wag right on seeing the owner, still right but less enthusiasm to a stranger, still right but definitely less with the cat, and a flip to the left upon seeing the aggressive dog.[5]

An elaboration of that study by the same group revealed that dogs watching videos (or even stills) of other dogs wagging to their left began to exhibit anxious behavior as their heart rates rose.[6]

Beagles in a similar situation, unlike the dogs in these studies, were more cautious about an unfamiliar person and initially, upon seeing them, wagged to the left. But after only five-minute sessions with the same person, for three days in a row, fifteen minutes total, the beagles were won over and were wagging hard to the right.[7] It's a signal that tells you something instantly about the dog's state of mind.

And other dogs know that. You'd have to believe that in addition to the tail, other dogs are paying attention to myriad signals, but we really only have scraps of knowledge about the intent of the signal sender.

For instance, how did it happen that the left-right movement of the tail is the cue to a dog's inner feelings? There is some indirect evidence that this is linked to the right and left hemispheres of the brain (as we saw with the perception of words). The left hemisphere controls the right side of the body and vice versa. So a tail wag to the right (contentment) is largely driven by the left hemisphere. And in humans, the left hemisphere is, to put it pretty crudely, a "happier" hemisphere—positive emotions are processed there, with negative on the right. This fits nicely with the tail-wagging story, but of course some more precise analysis of what's going on in the dog's brain would be pretty cool.

And why wagging? There's some speculative thinking on this,

The Future

based on the idea that because humans react positively to consistent, rhythmic movements (there's much evidence for this), that might have reinforced the back-and-forth of wagging that plays to humans.[8] But tail wagging didn't come out of nowhere as dogs were becoming domestic—wolves wag their tails too. But when wolves and dogs were raised together, dog pups began wagging their tails in the presence of their human at about four weeks; wolf pups did not. So tail wagging has flourished as dogs became closer to humans.

For creatures like us who lack tails, their deep potential for communication comes a distant second to vocalizations, but obviously, there is much more to the suite of communicative devices in both species. Can we tap into that whole system with the ever-increasing sophistication of artificial intelligence?

Every pet, if it thinks at all, thinks differently. Each communicates differently, and importantly, each would communicate differently with humans. This is, once again, where the millennia-long relationship with dogs and cats puts them front and center. There has been enough time for those animals to begin to evolve communication strategies specifically targeted at humans. There are many books on the market that go into great detail about the ways they have commandeered our communication channels—they are the only two pets well established enough in their relationships with humans to make it even remotely possible that using AI, we can better understand what's going on in their brains. Horses, pigs, and parrots might eventually fall into line, but when it comes to the Doctor Dolittle challenge, cats and dogs are the best place to start.

– CHAPTER 33 –

Future Pets

The dogs of science fiction include Krypto, Superman's fellow Kryptonian, who shares his X-ray vision and ability to fly; Blood, hero of *A Boy and His Dog*, who is telepathic and genetically altered; Muffit II, the robot dog from 1978's *Battlestar Galactica*; Dug, from the animated film *Up*, a golden retriever who wears a collar to be able to speak; and K-9 of *Doctor Who*, blasting lasers with its nose. Oddly enough, they all look like breeds we might see today, just with technological upgrades.

And there are cats: Lying Cat knows when people are lying, Spot wanders the stars in the USS *Enterprise*, and Robert Heinlein's Pixel not only learns to talk but also can pass through solid matter. The sci-fi cats haven't changed much in appearance either, but maintain their unearthly combination of devotion and cunning.

There are no cats or dogs with these extraordinary abilities, at least not yet, but there will definitely be changes coming to pets. But soon?

How different will the world of (living) pets look in, say, 2125, a century from now? Impossible to say with any certainty, but the guesses—and that's exactly what they are—run the gamut from cautious extrapolations of what we've seen in the past to fanciful, extravagant, and unlikely inventions.[1] Here's a possible example: A "biohacker" named Josie Zayner has started a company called the Los Angeles Project. The goal is to create unusual pets, using the genetic

The Future

technology CRISPR. They've already started experimenting with green fluorescent rabbits and have speculated they might, by manipulating the genetics of narwhals, take a shot at a unicorn. After all, says Zayner, "I think, as a human species, it's kind of our moral prerogative to level up animals." There's a dubious philosophical position.[2]

But speculating is fun, and even some of the more sobersided projections portray a pet world that is very different from today. It's worth starting from recent history, because some of the trends that are already underway are likely to continue, at least for a while, but not all.

I doubt anyone could have predicted the explosion we've seen in the number of officially registered breeds of dog. In the last twenty-five years, there have been fifty-six new dog breeds recognized by the American Kennel Club, the authoritative voice in North America on such things. Dogs like the Leonburger, the Entlebucher mountain dog, the Norwegian Lundehund, the Xoloitzcuintli, and fifty-two more. You might be familiar with some of them, and you might even be sure they've been around longer than that, but they've only become *officially* recognized since the year 2000.

They bring the total number of official dog breeds to 201 for the American Kennel Club and 356 recognized by the Fédération Cynologique Internationale, the International Canine Federation. They really are a special class, given that only a small percentage of these end up in shelters. On the other hand, in the USA, about six million dogs arrive in shelters every year. Two-thirds of those, roughly four million, are adopted. By contrast, more cats are admitted to shelters, but more are adopted. At best, shelters around the world are close to or exceeding capacity, and let's not forget that as much as two-thirds of the world's cats and dogs are not living full-time with humans.[3]

It's not just pressure on shelters that will influence pet futures. Dr. William Silversides at the Faculté de médecine vétérinaire (FMV), Université de Montréal, told me there is also a growing trend to discourage or even disallow the breeding of cats and dogs that, while satisfying some sort of public demand, degrade the health of the animal. Norway has banned the breeding of the Cavalier King Charles spaniel and restricted breeding of British bulldogs—the spaniel for common heart defects and eye and joint problems, the bulldogs because their shortened muzzles create serious issues with breathing and other health problems.

The Scottish fold cat, which carries a dominant mutation that causes the development of abnormal cartilage and bone, was attractive because its ears folded, making the cat look like an owl. But they have multiple problems with joint pain and degeneration. They are now prohibited in their homeland, Scotland.

Savannah cats, a cross between a domestic cat and a wild African serval, were created in 2001, and while they are hugely popular on Instagram (Stryker the cat has eight hundred thousand followers), it's suspected the servals involved are part of the international illegal pet trade. In the UK, it is now illegal to own a first-generation cross without a Dangerous Wild Animal license.

Some good news: Mutations for genetic diseases can be eliminated fairly easily now, because complete readouts of pets' genomes are now available. This creates an incentive for breeders to provide detailed genetics analyses of their animals, and as that becomes widespread, genetic flaws should gradually dwindle and eventually disappear.*

Unfortunately, the balance between new breeds, genetically improved versions of existing breeds, and discarded ones is impossible to predict. Also, these ups and downs are only a part of the human equation, which, of course, drives the popularity of all pets.

Some breeds, as we've seen, ride both the rise and fall of fashion, appearing in movies and other human-centered foibles. During the

---

* If you're interested, Labgenvet (www.labgenvet.ca) provides detailed pet genetic information.

The Future

pandemic, 20 percent of American households adopted a pet, but as the pandemic eased, and inflation rose, those numbers dropped.

And as the global middle class continues to grow, their love of pets will boost the number of pets. In China, pet ownership increased by more than 100 percent between 2014 and 2019, a combination of more money in people's pockets, changes in pet regulations, and a falling birth rate—fewer children means more money, more space, and an unfulfilled need to parent. But a greater percentage of that increasing number of pet owners in the future will be urban dwellers. How will that affect their pet choices? Will those pets be rescues or not?

Climate change will also be an issue. I already own a dog that hates hot weather. How will the inevitable climb in temperatures affect pet owners and their pets? There are already signs that wild animals are reacting, changing their bodies to better cope with the heat: Larger body parts, like birds' bills or mammals' ears, tails, and legs, help shed heat faster. This has been observed in Australian parrots and the masked shrew. Pets, of course, are somewhat insulated from the harshness of climate, but inevitably, it will affect them.

It is definitely too soon to speculate about the effects of climate change or other environmental changes, but there is research already underway that might make our pets live longer.

Despite the fact that cats generally live longer than dogs (thirteen to seventeen years versus ten to thirteen years), the records for longevity are not that different. The oldest cat ever was Creme Puff, who lived an incredible thirty-eight years and three days in the last decades of the twentieth century. This age has been verified as accurate. The oldest verified cat living as of this writing in 2025 is Flossie, born December 29, 1995. There are claims of older cats, but none has been verified.

The dog version of this story contains more intrigue. Bobi, a male purebred Rafeiro do Alentejo, died October 21, 2023, supposedly thirty-one years and 165 days old. Bobi was first acclaimed as the oldest ever by Guinness World Records, but doubts about the legitimacy of the claim surfaced (absence of reliable documentation), and eventually Bobi's claim was disallowed. This left Bluey, an Australian cattle dog, who died in November 1939, as the all-time record-holder at twenty-nine years, five months. The oldest dog still living as of 2025 is Spike the Chihuahua, said to have been born November 10, 1999.

These extraordinarily long-lived individuals are difficult to explain, but there is a research program that seeks to help us understand these and other facets of dog aging, called, yes, the Dog Aging Project. Headquartered at the University of Texas at El Paso and the University of Washington, it has recruited more than fifty thousand dogs, all of which were volunteered by their humans. This isn't just about dogs, though; they were chosen because dogs and humans share large parts of their genomes and also diseases, making dogs a better fit than rodents as an animal model for humans. Also, because these dogs are not lab animals, but are instead living in the same circumstances as their owners, their life experience mimics humans' more closely. And while rodents' very short lifespans have endeared them to scientists as lab animals, the scientists involved in the Dog Aging Project think that dogs' much longer lives are a worthwhile price to pay for a much better understanding of both dogs' and humans' lives.

This is a good opportunity to update old thinking about dog years versus human years. It used to be said that one dog year was equivalent to seven human years. A ten-year-old dog was "seventy," and record setters like Bluey and Spike an inhuman two hundred plus years. Considering the oldest human ever was 122, it's apparent that the 1:7 ratio, while it might seem reasonable in the early years, eventually breaks down. The newly revised estimate is that a dog's first year is worth about fifteen human years, the second worth nine, and every year thereafter four to five years. So the standard ten-year-old dog, once compared to a seventy-year-old human, is now understood to be closer to a sixty-year-old, and the outrageously old world-record dogs would be somewhere between 136 and 164, depending on whether four or five years is used as the interval after the first two. Still outrageous! It should also not be forgotten that the biggest dogs live significantly shorter lives than the smallest. The typical spread would be Bernese mountain dogs living, on average, seven years but Chihuahuas thirteen.

This fact itself is puzzling, because generally large species live longer than small ones. Think elephants versus mice. One suggestion has been that large dogs push the pedal to the metal to get to their maximum size. A study of seventy-four dog breeds published in 2013 showed that every additional two kilograms (four pounds) of weight subtracted one month off a dog's life expectancy. As the authors said: "Large dogs die young because they age quickly. . . . Their adult life unwinds in fast motion."[4]

The Dog Aging Project might shed more light on that phenomenon, but it has an additional feature that pushes it into wholly new territory. Some of the dogs in the program are receiving the drug rapamycin to see if it might extend their lives. Rapamycin is already used in humans in cancer treatment and prevention of transplanted organ rejection, but an unexpected bonus has been that it has been found to extend lives in a variety of organisms, from fruit flies to mice.

Given that mice are more closely related to us than fruit flies, the data from them is the most intriguing. Seemingly, whether mice are four, nine, or nineteen months old (and nineteen is pretty late in a mouse's life), their lives are dramatically extended by the drug, in some cases by 40 percent! And there appear to be no negative side effects. The Dog Aging Project will advance this research both by using dogs rather than mice and following those taking rapamycin for as long as they live. But imagine if a dog's average life is ten or eleven: Rapamycin could extend that by four or five years. The longest living dog might be thirty-five! And the future world of pets would look very different. The turnover of the dog population would be dramatically slowed.

A similar but smaller-scale project is also housed at the University of Washington, the Cat Healthy Aging Project. Like the Dog Aging

Project, it is built on a resemblance between cat and human diseases, especially Alzheimer's disease, in which cats develop brain deposits similar to those found in humans suffering from Alzheimer's—and unlike dogs. In this study, however, cats are not followed during their lives, but are donated for autopsy after they die. The study admits cats that have been euthanized because of disease. Their bodies are delivered to the lab, autopsied, and the cremated remains returned to the owners.[5]

Both the dog and cat studies should provide useful information about diseases common to all three species, but only the dog study might actually extend the lives of dogs, and maybe, given the disparity between the life expectancies of dogs and cats at the moment, will even them out. And you can be sure that proponents of extending human life expectancy will be watching closely.

We won't see any effects of the dog and cat projects for years, and life expectancy is only one feature of pets that might change significantly. The immediate future should generate more new breeds of dog, but at the same time, the number of mixed-breed animals will grow. But there's reason to think those mixed breeds, as a group, will head in a different direction: toward uniformity, rather than diversity.

As I've mentioned, the vast majority of the world's dogs and cats, many hundreds of millions of them, do not live in people's homes, and most attempts to curb their population growth have had limited success. If they're completely feral, they don't really qualify as pets, but they can certainly count pets in their ancestry. What's going to happen to them? In the absence of something like trap-neuter-return for cats (but a more effective technique), it's hard to project the future for either, but some scientists have taken a shot at imagining a crazy, hopefully-never-going-to-happen scenario in which, in the absence of humans, feral cats and dogs roam free. Call this an extreme thought experiment. Before considering that radical scenario, though, some thoughts on the immediate future.

While both dogs and cats are capable of hunting, I think a good case can be made that feral dogs are called "village dogs" for a good reason: They depend on handouts, scraps, and trash from humans. As the human population expands, this lifestyle will remain sustainable, unless the mass migration of people into cities that is already underway makes it harder to access these sources of food. After all, they're called village dogs, not city dogs, for a reason.

Jonathan Losos, a biologist at Washington University in St. Louis, has written a provocative article called "House Cats Will Rule the World."[6] Losos takes an evolutionary approach, starting with the fact that cats are already established on every continent except Antarctica. In so doing, they have encountered—and flourished in—an astonishing array of environments, from deserts to mountains to grasslands. Each unique environment presents a variety of challenges: where and how to access prey and shelter, how to avoid predators, how to adapt to seasonal change. Losos contends that the combination of such environmental pressures and cats' rapid reproduction suggests that they are already evolving.

The diversity of their habitats is also a factor because isolated populations tend to evolve in their own direction; islands are a perfect example. Desert cats will evolve differently from rainforest cats or prairie cats. Inevitably, but unpredictably, some populations will outstrip others, but regardless of success, diversity will be the result. Losos also points out the gloomy scenario that if we are indeed in the midst of the sixth great extinction—and there are very few scientists who doubt that we are—opportunities will open up for many species, as has happened in the past. Humans are largely responsible for the endangerment of predators like African wild dogs and the Philippine eagle, and when a predator is eliminated, another will take its place. If the prey is appropriate, feral cats can step in.*

---

* The irony in extinctions today is that feral cats are both (at least partly) agents of extinction and beneficiaries.

Future Pets

Losos ends with an extravagant thought. Remember saber-toothed cats? Could domestic cats eventually evolve into saber-tooths? He points out that "saber teeth evolved multiple times in cats and their relatives (and also once in South American marsupials). For much of cat history, saber-toothed species were more common than their less toothy relatives." Maybe they'll walk the earth again.

If they do, the descendants of today's feral dogs might choose to give them a wide berth. Village, or feral, dogs are usually brownish, not too big, short coat, narrow head, definitely not plump, pretty chill on the street. Average in many ways. Nonetheless, they are survivors, at least as a "breed"; individuals rarely reach seven years. Village dogs definitely don't enjoy the perks of pethood.

Marc Bekoff and Jessica Pierce took the idea of village dogs' estrangement from human caregivers one step further in their book *A Dog's World: Imagining the Lives of Dogs in a World Without Humans*.[7] They envisioned a future world where there are no humans on the planet and asked, "How would dogs do?" It's a great thought experiment, although necessarily they found it difficult to come to conclusions. It's just impossible to know. You'd think the biggest dogs would prevail, but maybe small dogs would—paradoxically—do better. They require less food, they might be less tempting to the larger predators, and they could find hiding places more readily.

Bekoff, in a piece written for *Psychology Today*, quotes one dog expert, Mark Derr, as suggesting that the village dogs of that future, human-free world would generally end up being pit bull hounds of roughly twenty-three to thirty-two kilograms (fifty to seventy pounds). Derr also suggested that feral dogs might form alliances with feral cats. The stuff of horror stories![8] Dan O'Neill, a veterinarian and a professor of animal epidemiology at Royal Veterinary College in the United Kingdom, said: "If humans disappeared off the planet tomorrow, breeds would be gone within five years because all the different types of dogs would breed with each other and morph back into one."[9]

The Future

The choke point in this imagined evolution of domestic dogs into "wild" dogs would be the ability to find enough food. No more garbage cans, roadkill, or landfills, just sparse hunting. But hunting skill would likely not be enough: Social skills—the ability to hunt cooperatively, band together for protection, and who knows what else—would be crucial. Any modern breed whose health is compromised by breeding (I'm looking at you, brachycephalics) would likely be quickly eliminated.*

It's true that today both cats and dogs can make it in a partially domestic, outskirts-of-town world, and a smaller number can go completely wild. At the moment, cats might be better tuned to living wild—a significant number continue to live on rodents and birds—but if humans were truly absent, those buffets of rats and mice would certainly drop, and cat life would be harder.

Cats might be pre-adapted, but there's one curious example of survival that suggests dogs might do well enough. In the immediate aftermath of the Chernobyl nuclear accident, deadly radiation concentrated at the wreckage of the reactor spread out on the prevailing winds, traveling thousands of kilometers. Humans were evacuated from the area, and efforts were made to cull all the dogs present, to prevent them from wandering and spreading the radiation they had absorbed. But the cull was incomplete, as they often are, and the dog population swelled, mixing original Chernobyl dogs that never left with newcomers.

It would have been reasonable to expect the dogs to die or accumulate a host of life-shortening fatal mutations, exposed as they were to a blast of radiation, but on the other hand, the remaining buildings offered shelter, and visiting humans provided both food and veterinary care. And there is evidence that some of them are turning more to their

---

* This includes pugs, bulldogs, shar-peis, and the Cavalier King Charles spaniel, among others.

own devices rather than depending on humans. There's no doubt that wildlife around Chernobyl, ranging from butterflies and voles to birds, has been affected by radiation. The genetic analysis isn't finished, but so far, it's clear that two separate groups of dogs, one living near the reactor and another about ten kilometers (a little over six miles) away, have startlingly different sets of mutations, the significance of which isn't yet clear. The remarkable survival of the Chernobyl dogs might say something about a future that feral dogs, in the absence of humans, would experience.[10]

One thing nobody studying Chernobyl's dogs is worried about? How much they'd miss human contact. Friederike Range, of the University of Veterinary Medicine in Vienna, said, "If we were to disappear, the food would be the main problem for the dogs, not losing the human as a social partner. As long as they could find food, they would be perfectly happy without us."[11]

A quick look around the world would argue that the risk of another nuclear wasteland is higher than that of the global extinction of humans—although they are definitely related—and so the Chernobyl studies are likely more "real-world" than the predictions about a world without any humans at all. But they both have something to say about the future of pets. It's satisfyingly consistent: Both dogs and cats, in the form of wildcats and wolves, were around before us, and it's just possible they'll be around after us.

Conclusion

We appear to be the *only* species on earth that keeps pets. More than that, we are so immersed in pet-keeping—our behaviors can be so entangled—that it's difficult to distinguish the animals' native behavior from what they've learned to curry favor with their humans. At the same time, the science aimed at understanding the pets *themselves* is pushing forward.

What I've taken away from writing this book is that the more you learn about pets, their histories and unique attributes, the richer your experiences with them. Now, when I meet a dog on the street, a cat in a home, or a lizard in an aquarium, I pay a lot more attention, trying to understand what they're up to or how they're feeling. I'll check the direction the dog's tail is wagging—if it is wagging at all. I try unsuccessfully to figure out what the cat in the chair is paying attention to. I'll stay out of the lizard's warmth-giving light. When you acknowledge that in any pet's head, there's a lot going on that you don't see or understand, your interactions might reveal something new: In the struggle to understand them, you get closer.

This, of course, applies to any pet, but dogs and cats have a leg up (or dogs do anyway). Remember how they got here: Dogs and cats were wild animals (some would argue cats retain aspects of this wildness) that became domesticated—that is, bred to live comfortably near or actually with human beings. Domestication was a turning point for us and the animals we chose. It's hard to find consistent dates, but all timelines agree that dogs were first; then pigs, sheep, and cattle all fall in line around nine or ten thousand years ago. Cattle follow, then chickens. Where are the cats? Some timelines I saw omitted cats completely (!), others suggested very different dates for their origin, but taking it all together, cats were likely either the second or third animal to be domesticated, either after dogs, or after dogs and either sheep, pigs, or goats.

So our favorite pets—dogs and cats—are two of the three earliest animals to be domesticated. The most popular pets have had the longest time with us to create their own version of a relationship with a human.

Much as it looks like pet-keeping has suddenly become a global obsession, remember that thousands of years ago, dogs and cats were kept in luxurious conditions (some of them anyway), being pampered by royalty the same way they are now by commoners. They were then, as now, accessories, companions, and even employees. Other than the rise and fall in popularity of some pets as the centuries have passed, there is little to choose between the kind of pet care that was provided in ancient Egypt and now.

Even as we pay such close attention to pets, we continue to evolve together. We ask, as their human companions, how can we serve them best (although, to be honest, many of our decisions are based on how they could serve us better). Much of the owner-pet relationship might seem straightforward, but the bewildering array of pets combined with multiple shades of human attitudes toward them means we have made only slight progress toward untangling that relationship.

And so, despite the apparent consistency in pet-keeping over millennia, might we be at some kind of turning point? As more humans become urban dwellers, and the relationship to wild environments diminishes, what new benefits in a mutual dependency might evolve? Will the average size of a pet (and therefore choice of breed or species)

Conclusion

decline in response to shrinking living space? Will a reduction in the exposure to nature create changes in our behavior toward them?

We are also heading toward—if not already in—an environmental crisis, where the impact of pets on wildlife and the production of pet food on the environment will surely have to change. Such changes will inevitably challenge the balance between pet owners' rights and responsibilities. And to the pets themselves . . . will the current desire to unlock some of the mysteries of the pet "mind" result in new insights that change the way we think of pets, and ourselves?

It's about the depth of our relationships with other animals, that biophilia. It's most intense with pets, but it connects us to wildlife as well. On the one hand, there is the desire of some pet owners to extend their pet's presence after death as a way of dealing with loss. On the other, there's the uncomfortable realization that wildlife is in decline. Anyone who has spent any time in nature over the last few decades knows there are species that you used to see often that now have dwindled dramatically. I encountered whip-poor-wills routinely every summer at my place in Ontario. Actually, you'd hear them more than see them, but they were a predictable feature of summer evenings. Not anymore. I haven't heard a whip-poor-will for a couple of years now, probably more. It's not just me or my particular location: Their populations have been shrinking for at least fifty years.

I might never see or hear a whip-poor-will again. That is a deep disappointment, even though I've never communicated with one. While the communication is important—it leads to a stronger attachment—even in the absence of it, we can form attachments that hurt when they're broken.

What's to become of those time-honored ways of thinking about pets if indeed we start to probe much deeper into their minds, only to find that there isn't much there? Or, conversely, that the exuberant activity we find in their brains is indecipherable? Given the consistency with which we have treated some pets over the centuries, it might make sense to predict that nothing much would be different. However, there is the risk that gaining insights into your dog or cat's thinking could reveal that they're really not as devoted to you as you think. Or hope.

But whatever we learn, it will likely give us a deeper understanding of them and an even greater attachment. I hope some of that will spill over to wild populations, helping to establish some sort of balance between the growing human population, our pets, and the wild environment and its inhabitants. We could face a future in which pets *are* surrogates for the wildlife we once knew; in which instead of having relationships with other animals, we become only keepers of animals. Thankfully, pet science suggests we can evolve in the other direction—to greater communication and respect.

And better understanding. Why are we the only species on earth that keeps pets (although, keeping ancient hominids in mind, we really should say the only *living* species)? We have no answer for that—at least now.

Many questions need answers. The last words go to author and geographer Yi-Fu Tuan, who, while commenting on dog ownership, made a point that surely can be made about any pet, today or in the future: that they can encourage "on the one hand the best that a human person is capable of—self-sacrificing devotion to a weaker and dependent being, and, on the other hand, the temptation to exercise power in a willful and arbitrary, even perverse, manner. Both traits can exist in the same person."[1]

Acknowledgments

I am grateful to the crew at Simon & Schuster Canada for shepherding this book through to publication. I don't know if this is common, but Jim Gifford, who edited some of my books years and years ago, suddenly reappeared to edit this one. The odd thing is that he seems to have had more curious experiences with unusual pets than I have. However, it didn't seem appropriate to interview him for a book he was editing.

I'm indebted to Kevin Hanson for the inspiration for the book and to those behind-the-scenes people responsible for publicity, cover art, social media marketing, and innumerable things I'm not even aware are happening. So a big thank-you to the president and publisher of S&S Canada, Nicole Winstanley (not so much behind-the-scenes!), my publicist Lisa Wray, Muna Hussein, Kaitlyn Lonnee, Sarah Lachmansingh, Michael Guy-Haddock, and Dan French. And what's a book without an agent to guide it? My thanks always to Jackie Kaiser et al. at Westwood Creative Artists.

The following experts contributed indirectly. They were among the interviewees who appeared on the *Anthropomania* podcast, and their ideas played an important role here. They are Hal Herzog, James Serpell, Suzanne MacDonald, Stan Coren, and Lisa Cooper.

In the course of writing the book, I had important conversations with Chris Darimont, Carolyn Willekes, Heather Clitheroe, Liz Murray,

Acknowledgments

Shukoo Nasim, Aaron Fairweather (Entobird), Allis Markham, Celina Juliano, Jennifer Vonk, Sam Gosling, Seonaid Eggett, Lisa Carver, William Silversides, Kate Creevy, and Matt Kaeberlein.

Of the scientists who have devoted time and energy both to pets—and, more precisely, the human relationship to animals—I'd recommend anything written by James Serpell and Hal Herzog, but specifically Serpell's *In the Company of Animals* and Herzog's *Some We Love, Some We Hate, Some We Eat*. Also Yi-Fu Tuan's unsettling *Dominance and Affection: The Making of Pets* for context on pet-keeping. If you want to drill deeper into dogs and cats, I'd recommend anything by Alexandra Horowitz on dogs and John Bradshaw on cats. I also acknowledge there are likely hundreds of other pet books worth reading.

Not that I didn't appreciate the pets I've had before diving into the research, but I feel I understand them better. However, I have to admit that I still puzzle over the behavior of my current pet, Robbie the dog, especially his behavioral interactions with his best canine friend, Penny. They're both trying to tell us something.

Notes

INTRODUCTION

1. "Why PETA Wants You to Stop Saying 'Pet,'" PETA UK, February 4, 2020, https://www.peta.org.uk/blog/peta-pets/.
2. Yi-Fu Tuan, *Dominance and Affection: The Making of Pets* (Yale University Press, 1984).

CHAPTER 1: BIOPHILIA

1. E. O. Wilson, *Biophilia* (Harvard University Press, 1984).
2. E. O. Wilson and S. Kellert, *The Biophilia Hypothesis* (Island Press, 1993).
3. "A Conversation with E. O. Wilson," *Nova*, PBS, April 1, 2008, https://www.pbs.org/wgbh/nova/article/conversation-eo-wilson/.
4. E. O. Wilson, "Behavior of the Cuban Lizard *Chamaeleolis chamaeleontides* (Duméril and Bibron) in Captivity," *Copeia* 2 (1957).
5. Doug Matthews, "The Grand Procession of Ptolemy Philadelphus: Part Two," *Ancient Celebration* (blog), March 24, 2011, https://ancientcelebration.blogspot.com/2011/03/grand-procession-of-ptolemy_24.html.
6. Rohini Mohan, "Critics Not Wild over Private Zoo Owned by Indian Billionaire Ambani's Youngest Son," *The Straits Times*, March 17, 2024, https://www.straitstimes.com/asia/south-asia/critics-not-wild-over-private-zoo-owned-by-indian-billionaire-ambani-s-youngest-son.
7. Teresa Telecky and Doris Lin, "Big Game, Big Bucks: The Alarming Growth of the American Trophy Hunting Industry," *Big-Game and Trophy Hunting Collection* 7 (1995).
8. Wilson, *Biophilia*.
9. Lisa Mayor, "'Hunting Is Not About Killing for Me': Trophy Hunter Sees Shooting Big Game as Form of Conservation," CBC News, February 10, 2017, https://www.cbc.ca/news/canada/trophy-hunting-jacine-jadresko-1.3974716.
10. Wilson, *Biophilia*.
11. C. Amiot, C. Gagne, and B. Bastian, "Exploring the Role of Our Contacts with Pets in Broadening Concerns for Animals, Nature, and Fellow Humans: A Representative Study," *Scientific Reports* 13 (2023): 17079.

CHAPTER 2: WHAT IS A PET?

1. Kasey Grier, "What Is a Pet?," *The Pet Historian* (blog), accessed April 14, 2025, https://thepethistorian.com/what-is-a-pet/.
2. Grier, "What Is a Pet?"
3. John Bradshaw, *The Animals Among Us: The New Science of Anthrozoology* (Basic Books, 2017).
4. Timothy J. Eddy, "What Is a Pet?" *Anthrozoös* 16, no. 2 (2003): 98.
5. Anna Brown, "About Half of U.S. Pet Owners Say Their Pets Are as Much a Part of Their Family as a Human Member," Pew Research Center, July 7, 2023, https://www.pewresearch.org/short-reads/2023/07/07/.

CHAPTER 3: WHEN DID IT ALL START?

1. S. J. M. Davis, "Evidence for Domestication of the Dog 12,000 Years Ago in the Natufian of Israel," *Nature* 276 (December 7, 1978): 608–610.
2. Loretta Cormier, "Animism, Cannibalism and Pet-Keeping Among the Guajá of Eastern Amazonia," *Tipití: Journal of the Society for the Anthropology of Lowland South America* 1, no. 1 (2003).
3. William Ellis, *Polynesian Researches*, vol. 1, 2nd ed. (Fisher, Son & Jackson, 1841).
4. Veronika Simonova, "The Wild at Home and the Magic of Contact: Stories About Wild Animals and Spirits from Amudisy Evenki Hunters and Reindeer Herders," *Études mongoles et sibériennes, centrasiatiques et tibétaines* 49 (2018).
5. Simonova, "Wild at Home and the Magic of Contact."
6. James Serpell, "Pet-Keeping in Non-Western Societies: Some Popular Misconceptions," *Anthrozoös* 1, no. 3 (Winter 1987).

CHAPTER 4: WHY DO WE KEEP PETS?

1. J. New, L. Cosmides, and J. Tooby, "Category-Specific Attention for Animals Reflects Ancestral Priorities, Not Expertise," *Proceedings of the National Academy of Sciences* 104, no. 42 (October 16, 2007): 16598–16603.
2. Yann Leroux, "Experimental Study of Apparent Behavior. Fritz Heider & Marianne Simmel. 1944," YouTube video, December 26, 2010, https://www.youtube.com/watch?v=n9TWwG4SFWQ.
3. "Out of Shape," USC ICT, YouTube video, February 12, 2014, https://www.youtube.com/watch?v=ZAnt9II-5Co.
4. E. O. Wilson, *Biophilia* (Harvard University Press, 1984).
5. Charles Darwin, *The Descent of Man, and Selection in Relation to Sex*, vol. 1 (John Murray, 1871).
6. Thomas Nagel, "What Is It Like to Be a Bat?," *Philosophical Review* 83, no. 4 (October 1974): 435–450.

CHAPTER 5: DOES PET-KEEPING EVEN MAKE SENSE?

1. M. Glocker et al., "Baby Schema in Infant Faces Induces Cuteness Perception and Motivation for Caretaking in Adults," *Ethology* 115, no. 3 (2009): 257–263.
2. Stephen Jay Gould, "Mickey Mouse Meets Konrad Lorenz," *Natural History* 88, no. 5 (May 1979): 30–36.
3. Kathryn Cyr and Roxanne Hawkins, "'Can't Imagine Having One Without the Other Now': Maternal Perceptions of Pets and Attachment Relationships in the Perinatal Period," *Human-Animal Interactions* 13, no. 1 (January 31, 2025): 1–9.
4. Francis Galton, "The First Steps Towards the Domestication of Animals," *Transactions of the Ethnological Society of London* 3 (1865): 122–138.
5. James Serpell, *In the Company of Animals* (Cambridge University Press, 1996).
6. John Archer, "Why Do People Love Their Pets?" *Evolution and Human Behavior* 18 (1997): 237–259.

CHAPTER 6: WHERE DID DOGS COME FROM?

1. Wikipedia, "Paleolithic dog," (last modified December 25, 2024), https://en.wikipedia.org/wiki/Paleolithic_dog.
2. Francis Galton, "The First Steps Towards the Domestication of Animals," *Transactions of the Ethnological Society of London* 3 (1865): 122–138.

CHAPTER 7: WHERE DID CATS COME FROM?

1. W. M. Conway, "The Cats of Ancient Egypt," in *The English Illustrated Magazine*, vol. 7 (Macmillan, 1890), 253, https://archive.org/details/englishillustra27unkngoog/page/253/mode/2up.
2. A. R. Williams, "Animals Everlasting," *National Geographic* (November 2009).
3. James Serpell, "Domestication and History of the Cat," in *The Domestic Cat*, ed. Dennis C. Turner and Patrick Bateson (Cambridge University Press, 2013), 97. It's also true that the modern phrase "there's more than one way to skin a cat" bore some unfortunate truth in medieval Spain. Some nine hundred domestic cat bones were discovered in a pit. Cut marks on the bones suggested the animals were skinned for their fur.
4. Irina Metzler, "Heretical Cats: Animal Symbolism in Religious Discourse," *Medium Aevum Quotidianum* 59 (2009): 4.
5. Spencer McDaniel, "Were Cats Really Killed En Masse During the Middle Ages?," *Tales of Times Forgotten* (blog), November 5, 2019, https://talesoftimesforgotten.com/2019/11/05/were-cats-really-killed-en-masse-during-the-middle-ages/.
6. G. Elvers et al., "Explicit and Implicit Measures of Black Cat Bias in Cat and Dog People," *Animals* 14 (2024): 3372.

7. P. Pongracz and C. A. Lugosi, "Predator for Hire: The Curious Case of Man's Best Independent Friend, the Cat," *Applied Animal Behaviour Science* 271 (February 2024).
8. John Bradshaw, *Cat Sense* (Basic Books, 2014).

CHAPTER 8: COSTS/BENEFITS TO HUMANS

1. Pat Shipman, "The Animal Connection and Human Evolution," *Current Anthropology* 51, no. 4 (August 2010): 519–538.
2. Michelle Megna, "Pet Ownership Statistics 2025," *Forbes*, January 2, 2025, https://www.forbes.com/advisor/pet-insurance/pet-ownership-statistics/.
3. Hal Herzog, "The Sad Truth About Pet Ownership and Depression," *Psychology Today*, December 2019, https://www.psychologytoday.com/us/blog/animals-and-us/201912/the-sad-truth-about-pet-ownership-and-depression.
4. E. Friedmann et al., "Animal Companions and One-Year Survival of Patients After Discharge from a Coronary Care Unit," *Public Health Reports* 95, no. 4 (July–August 1980): 307–312.
5. E. Friedmann et al., "Pet's Presence and Owner's Blood Pressures During the Daily Lives of Pet Owners with Pre- to Mild Hypertension," *Anthrozoös* 26, no. 4 (2013): 535–550.
6. G. Levine et al., "Pet Ownership and Cardiovascular Risk: A Scientific Statement from the American Heart Association," *Circulation* 127 (2013): 2353–2363.
7. B. Kretzlet et al., "Pet Ownership, Loneliness, and Social Isolation: A Systematic Review," *Social Psychiatry and Psychiatric Epidemiology* 57 (2022): 1935–1957.
8. M. K. Mueller et al., "Companion Animal Relationships and Adolescent Loneliness during COVID-19," *Animals* 11 (2021): 885.
9. Nik Taylor and Tani Signal, "Empathy and Attitudes to Animals," *Anthrozoös* 18, no. 1 (2005): 18–27.
10. Nicolas Guéguen and Serge Ciccotti, "Domestic Dogs as Facilitators in Social Interaction: An Evaluation of Helping and Courtship Behaviors," *Anthrozoös* 21, no. 4 (2008): 333–349.
11. Deborah L. Wells, "The Facilitation of Social Interaction by Domestic Dogs," *Anthrozoös* 17, no. 4 (2004): 340–352.

CHAPTER 9: COSTS/BENEFITS TO PETS

1. Matthew Gompper, "The Dog-Human-Wildlife Interface: Assessing the Scope of the Problem," in *Free-Ranging Dogs and Wildlife Conservation* (Oxford University Press, 2014).
2. J. Hughes and D. MacDonald, "A Review of the Interactions Between Free-Roaming Domestic Dogs and Wildlife," *Biological Conservation* 157 (2013): 341–351.
3. Andrew Rowan, "Global Dog Populations," WellBeing International, May 30, 2020, https://wellbeingintl.org/global-dog-populations-2/.

4. Francis Galton, "The First Steps Towards the Domestication of Animals," *Transactions of the Ethnological Society of London* 3 (1865): 122–138.
5. Peter Gray and Sharon Young, "Human-Pet Dynamics in Cross Cultural Perspective," *Anthrozoös* 24, no. 1 (2011): 17–30.
6. Merrill Singer, "Pygmies and Their Dogs: A Note on Culturally Constituted Defence Mechanisms," *Ethos* 6, no. 4 (Winter 1978): 270–277.

CHAPTER 10: HORSES

1. Interview with Heather Clitheroe.
2. Jane Flynn, *Soldiers and Their Horses: Sense, Sentimentality and the Soldier-Horse Relationship in the Great War* (Routledge, 2020).
3. "The Pitiable Martyrdom of Man's Faithful Friend," *War Illustrated*, October 3, 1914.
4. Jane Flynn, "Goodbye Old Man? The Evolution of the Soldier-Horse Relationship in Myth and Memory, 1880–1939," *Humanimalia* 14, no. 2 (Spring 2024).
5. Elizabeth Banicki, "What Do Horses Feel at the Kentucky Derby? Mostly Fear and Pain," *The Guardian*, May 3, 2023.
6. A. Wisniewska et al., "Heterospecific Fear and Avoidance Behaviour in Domestic Horses (*Equus caballus*)," *Animals* 11 (2021): 3081.
7. I. Janczarek et al., "Social Behaviour of Horses in Response to Vocalisations of Predators," *Animals* 10 (2020): 2331.
8. Temple Grandin and Catherine Johnson, *Animals in Translation: Using the Mysteries of Autism to Decode Animal Behavior* (Harcourt, 2005).
9. C. Dubois et al., "Examining Canadian Equine Industry Participants' Perceptions of Horses and Their Welfare," *Animals* 8 (2018): 201.
10. Louise Evans et al., "Whoa, No-Go: Evidence Consistent with Model-Based Strategy Use in Horses During an Inhibitory Task," *Applied Animal Behaviour Science* 277 (2024): 106339.
11. Judith Tarr, "Transcending Words: The Real Fantasy of Human-Animal Communication," *Reactor Magazine*, April 2022, https://reactormag.com/transcending-words-the-real-fantasy-of-human-animal-communication/.
12. Judith Tarr, "Understanding and Writing Horses: Interspecies Communication," *Reactor Magazine*, February 2021, https://reactormag.com/understanding-and-writing-horses-interspecies-communication/.

CHAPTER 11: PARROTS

1. "The Truth About Parrots as Pets," In Defense of Animals, accessed April 14, 2025, https://www.idausa.org/campaign/wild-animals-and-habitats/parrots-as-pets/.

2. Karen Trinkaus, "The Pros and Cons of Owning Parrots," *Feisty Feathers* (blog), March 14, 2016, https://feistyfeathers.com/2016/03/14/the-pros-and-cons-of-owning-parrots/.
3. T. J. S. Balsby, J. V. Momberg, and T. Dabelsteen, "Vocal Imitation in Parrots Allows Addressing of Specific Individuals in a Dynamic Communication Network," *PLOS One* 7, no. 11 (2012).
4. E. Colbert-White et al., "Social Context Influences the Vocalizations of a Home-Raised African Grey Parrot (*Psittacus erithacus erithacus*)," *Journal of Comparative Psychology* 125, no. 2 (2011): 175–184.
5. "The Moth: Alex & Me—Irene Pepperberg," World Science Festival, YouTube video, December 5, 2013, https://www.youtube.com/watch?v=rrX1nrvPbLY.
6. Michelle Starr, "These Parrots Show Kindness to Each Other Without Any Benefit to Themselves," *ScienceAlert*, January 10, 2020, https://www.sciencealert.com/african-grey-parrots-show-even-the-bird-brained-like-to-help-their-friends.
7. E. Colbert-White et al., "Compositional Differences, Functional Similarities: A Linguistic Analysis Private Speech from a Young Child and a Home-Reared African Grey Parrot (*Psittacus erithacus erithacus*)," *International Journal of Comparative Psychology* 37 (2024): 2–19.
8. T. Roubalová et al., "Comparing the Productive Vocabularies of Grey Parrots (*Psittacus erithacus*) and Young Children," *Animal Cognition* 27, no. 45 (2024).

CHAPTER 12: ANTS

1. Stephen R. Kellert, "Values and Perceptions of Invertebrates," *Conservation Biology* 7, no. 4 (December 1993): 845–855.
2. Joan Embery and Ed Lucaire, *Joan Embery's Collection of Amazing Animal Facts* (Dell, 1983).
3. F. Ratnieks and T. Wenseleers, "Altruism in Insect Societies and Beyond: Voluntary or Enforced?" *Trends in Ecology and Evolution* 23 (2008): 45–52.
4. Vikram Chandra et al., "Social Regulation of Insulin Signaling and the Evolution of Eusociality in Ants," *Science* 361, no. 6400 (2018): 398–402.
5. Adrian Smith et al., "Cuticular Hydrocarbons Reliably Identify Cheaters and Allow Enforcement of Altruism in a Social Insect," *Current Biology* 19 (2009): 78–81.

CHAPTER 13: HYDRAS

1. Original: A. Trembley, *Mémoires pour servir à l'histoire d'un genre de polypes d'eau douce, à bras en forme de cornes* (Jean and Hermann Verbeek, 1744). Translation: S. G. Lenhoff and H. M. Lenhoff, *Hydra and the Birth of Experimental Biology, 1744: Abraham Trembley's Memoires Concerning the Polyps* (Boxwood Press, 1986).

2. A. Karabulut et al., "The Architecture and Operating Mechanism of a Cnidarian Stinging Organelle," *Nature Communications* 13 (2022).

CHAPTER 14: PET PEOPLE

1. Susan Krause Whitbourne, "Just How Different Are Cat People and Dog People?," *Psychology Today*, April 2016, https://www.psychologytoday.com/intl/blog/fulfillment-any-age/201604/just-how-different-are-cat-people-and-dog-people.
2. L. R. Finka et al., "Owner Personality and the Wellbeing of Their Cats Share Parallels with the Parent-Child Relationship," *PLOS One* 14, no. 2 (2019).
3. Stan Coren, "Personality Differences Between Dog and Cat Owners," *Psychology Today*, February 17, 2010, https://www.psychologytoday.com/ca/blog/canine-corner/201002/personality-differences-between-dog-and-cat-owners.
4. Joy Aschenbach, "Grumpy to Gleeful; Bears at Play," *Los Angeles Times*, May 7, 1995, https://www.latimes.com/archives/la-xpm-1995-05-07-me-63341-story.html.
5. Lisa A. Curb et al., "The Relationship Between Personality Match and Pet Satisfaction Among Dog Owners," *Anthrozoös* 26, no. 3 (2013): 395–404.
6. Y. Bender et al., "What Makes a Good Dog-Owner Team? A Systematic Review About Compatibility in Personality and Attachment," *Applied Animal Behaviour Science* 260 (2023): 105857.

CHAPTER 15: PET NAMES

1. Kathleen Walker-Meikle, *Medieval Pets* (Boydell Press, 2012).
2. William Safire, "Name That Dog," On Language, *New York Times*, December 22, 1985, https://timesmachine.nytimes.com/timesmachine/1985/12/22/234826.html?pageNumber=310.
3. Thom Nelson and Mimi Padmabandu, "Most Popular Dog Names of 2024," Embark, February 2, 2024, https://embarkvet.com/resources/most-popular-dog-names/.
4. "Camp Bow Wow's Dog Name Trends for 2024," Camp Bow Wow, March 1, 2024, https://www.campbowwow.com/blogs/2024/march/camp-bow-wows-dog-name-trends-for-2024/.
5. "Most Popular Dog Names," Rover, November 2024, https://www.rover.com/blog/dog-names/.
6. "The Most Unusual Dog Names: What They Reveal About Them and You," Naturo Dog Food, July 2023, https://www.naturopetfoods.ie/blog/advice/revealing-the-psychology-behind-unusual-dog-names.
7. "More Than a Pet," Humane World for Animals, accessed April 14, 2025, https://www.humaneworld.org/en/campaign/more-than-pet.
8. Alexa Albert, "The Significance of Names," *Anthrozoös* 1, no. 3 (1987).

CHAPTER 16: THE WHIMSY OF DOG BREEDS

1. Jacob Osborn, "Every New Dog Breed Recognized in the 21st Century," *Stacker*, January 22, 2024, https://stacker.com/pets/every-new-dog-breed-recognized-21st-century.
2. Sinding et al., "Arctic-Adapted Dogs Emerged at the Pleistocene-Holocene Transition," *Science* 368, no. 6498 (2020): 1495–1499.
3. Natalia Borecka, "100 Years of Canine Couture: Decade Defining Dog Breeds Through the Lens of Fashion," *Lone Wolf Magazine*, https://lonewolfmag.com/canine-couture-decade-defining-dog-breeds-lens-fashion/.
4. Hilton herself was blamed for Chihuahuas overwhelming animal shelters in California in 2010 as her fans emulated her and her dog Tinkerbell by purchasing a Chihuahua . . . and then changing their minds.
5. H. Herzog, "Biology, Culture, and the Origins of Pet-Keeping," *Animal Behavior and Cognition* 1, no. 3 (2014): 296–308.
6. S. Weir and S. E. Kessler, "The Making of a (Dog) Movie Star: The Effect of the Portrayal of Dogs in Movies on Breed Registrations in the United States," *PLOS One* 17, no. 1 (2022).
7. S. Ghirlanda et al., "Fashion vs. Function in Cultural Evolution: The Case of Dog Breed Popularity," *PLOS One* 8, no. 9 (2013).
8. P. Eretova et al., "Can My Human Read My Flat Face? The Curious Case of Understanding the Contextual Cues of Extremely Brachycephalic Dogs," *Applied Animal Behaviour Science* 270 (2024).
9. Alexandra Horowitz, "Has Dog Breeding Gone Too Far?," *New York Times*, May 9, 2024, https://www.nytimes.com/interactive/2024/05/09/opinion/purebred-dogs-inbreeding.html.
10. "Insurance Premiums for Brachy Breeds Are Highest," *Vet Record* 187, no. 1 (July 2020): 4–5, https://doi.org/10.1136/vr.m2756.
11. Amanda Fortini, "The Poodle Partying with the Kardashians and Cher," *The New Yorker*, March 11, 2024, https://www.newyorker.com/magazine/2024/03/18/the-poodle-partying-with-the-kardashians-and-cher.
12. Jan Bondeson, *Amazing Dogs: A Cabinet of Canine Curiousities* (Cornell University Press, 2011).

CHAPTER 17: EXOTIC PETS—EXOTIC PEOPLE?

1. A. Hergovich et al., "Exotic Animal Companions and the Personalities of Their Owners," *Anthrozoös* 24, no. 3 (2011).
2. M. Beverland et al., "Exploring the Dark Side of Pet Ownership: Status- and Control-Based Pet Consumption," *Journal of Business Research* 61, no. 5 (2008): 490–496.
3. "Exotic," *Merriam Webster's Collegiate Dictionary*, 11th ed. (2020).
4. Back et al., "Narcissistic Admiration and Rivalry: Disentangling the Bright and Dark Sides of Narcissism," *Journal of Personality and Social Psychology* 105, no. 6 (2013).

5. Jennifer Vonk, Chelsea Patton, and Moriah Galvan, "Not So Cold-Blooded: Narcissistic and Borderline Personality Traits Predict Attachment to Traditional and Non-Traditional Pets," *Anthrozoös* 29, no. 4 (2016): 627–637.
6. V. L. O'Connor and J. Vonk, "A (Tiger) King's Ransom: Dark Personality Features Predict Endorsement of Exotic Animal Exploitation," *Personality and Individual Differences* 202 (2023): 111956.
7. Melissa A. Smith, "What to Say to People Who Are Against Exotic Pet Ownership," HubPages, April 7, 2023, https://discover.hubpages.com/animals/simplelogic.
8. A. Hausmann et al., "Assessing Preferences and Motivations for Owning Exotic Pets: Care Matters," *Biological Conservation* 281 (2023): 110007.

CHAPTER 18: THE OUTDOOR CAT

1. Edward Howe Forbush, *The Domestic Cat: Bird Killer, Mouser and Destroyer of Wild Life; Means of Utilizing and Controlling It* (CreateSpace, 2018).
2. Cheryl Abbate, "A Defense of Free-Roaming Cats from a Hedonist Account of Feline Well-Being," *Acta Analytica* 34, no. 4 (2019): 1–23.
3. Bernard Rollin, "Telos," in *Veterinary and Animal Ethics: Proceedings of the First International Conference on Veterinary and Animal Ethics*, ed. Christopher M. Wathes et al. (Wiley, September 2011).
4. William G. George, "Domestic Cats as Predators and Factors in Winter Shortages of Raptor Prey," *Wilson Bulletin* 86, no. 4 (1974).
5. R. Kays and A. DeWan, "Ecological Impact of Inside/Outside House Cats Around a Suburban Nature Preserve," *Animal Conservation* 7 (2004): 1–11; K. A. Loyd et al., "Quantifying Free-Roaming Domestic Cat Predation Using Animal-Borne Video Cameras," *Biological Conservation* 160 (2013): 183–189.
6. R. Kays et al., "The Small Home Ranges and Large Local Ecological Impacts of Pet Cats," *Animal Conservation* 23, no. 5 (March 2020).
7. D. J. Herrera et al., "Prey Selection and Predation Behavior of Free-Roaming Domestic Cats (*Felis catus*) in an Urban Ecosystem: Implications for Urban Cat Management," *Biological Conservation* 268 (2022).
8. S. Loss et al., "The Impact of Free-Ranging Domestic Cats on Wildlife of the United States," *Nature Communications* 4 (2013): 1396.
9. Francis Battista, "Fuzzy Math on Cats, Birds Clouds Highly Questionable 'Study,'" Best Friends Animal Society, January 31, 2013, https://bestfriends.org/stories/julie-castle-blog/fuzzy-math-cats-birds-clouds-highly-questionable-study.
10. C. Y. Lepczyk et al., "A Global Synthesis and Assessment of Free-Ranging Domestic Cat Diet," *Nature Communications* 14 (2023): 7809.

11. "Misinterpretation of a Recent Study Threatens Cats' Lives," Alley Cat Allies, January 2024, https://www.alleycat.org/misinterpretation-of-a-recent-study-threatens-cats-lives/.
12. T. Doherty et al., "Invasive Predators and Global Biodiversity Loss," *PNAS* 113, no. 40 (2016): 11261–11265.
13. Patrick Pester, "Mice on Remote Island That Eat Albatrosses Alive Sentenced to Death by 'Bombing,' Scientists Decree," *Live Science*, August 29, 2024, https://www.livescience.com/animals/birds/mice-on-remote-island-that-eat-albatrosses-alive-sentenced-to-death-by-bombing-scientists-decree.
14. David Jessup, "The Welfare of Feral Cats and Wildlife," *JAVMA* 225, no. 9 (2004): 1377–1383.
15. A. Trouwborst et al., "Domestic Cats and Their Impacts on Biodiversity: A Blind Spot in the Application of Nature Conservation Law," *People and Nature* 2 (2020): 235–250.

CHAPTER 19: BEYOND CATS

1. Michael Taborsky, "Kiwis and Dog Predation: Observations in Waitangi State Forest," *Notornis* 35 (1988): 197–202.
2. J. A. Hernandez et al., "Dog Overpopulation on Santa Cruz Island, Galapagos 2018," *Conservation Science and Practice* 2, no. 6 (2020).
3. T. S. Doherty et al., "The Global Impacts of Domestic Dogs on Threatened Vertebrates," *Biological Conservation* 210 (2017): 556–559.
4. L. Bridson, "Minimising Visitor Impacts on Threatened Shorebirds and Their Habitats," New Zealand Department of Conservation (2000).
5. K. J. H. Williams, "Birds and Beaches, Dogs and Leashes: Dog Owners' Sense of Obligation to Leash Dogs on Beaches in Victoria, Australia," *Human Dimensions of Wildlife* 14, no. 2 (2024): 89–101.
6. G. S. Okin, "Environmental Impacts of Food Consumption by Dogs and Cats," *PLOS One* 12, no. 8 (2017).
7. Antonio Florio, "The First Recorded Python in Everglades National Park, 40 Years Later," National Parks System, October 2019, https://www.nps.gov/articles/the-first-recorded-python-in-everglades-national-park-40-years-later.htm.
8. Wolf-Christian Saul et al., "Assessing Patterns in Introduction Pathways of Alien Species by Linking Major Invasion Data Bases," *Journal of Applied Ecology* 54 (2017): 657–669.
9. Zoe Gentes, "Exotic Pets Can Become Pests with Risk of Invasion," Ecological Society of America, June 2, 2019, https://www.esa.org/blog/2019/06/03/exotic-pets-can-become-pests-with-risk-of-invasion/.
10. J. Lockwood et al., "When Pets Become Pests: The Role of the Exotic Pet Trade in Producing Invasive Vertebrate Animals," *Frontiers in Ecology and the Environment* 17, no. 6 (2019): 323–330.

11. Jérôme M. W. Gippet and Cleo Bertelsmeie, "Invasiveness Is Linked to Greater Commercial Success in the Global Pet Trade," *PNAS* 118, no. 14 (2021): e2016337118.
12. Gary Francione and Anna Charlton, "The Case Against Pets," *Aeon*, September 2016, https://aeon.co/essays/why-keeping-a-pet-is-fundamentally-unethical.
13. Troy Vettese, "Want to Truly Have Empathy for Animals? Stop Owning Pets," *The Guardian*, February 4, 2013, https://www.theguardian.com/commentisfree/2023/feb/04/want-to-truly-have-empathy-for-animals-stop-owning-pets.

CHAPTER 20: EAT . . . OR BE EATEN

1. Jonathan Haidt et al., "Affect, Culture and Morality, or Is It Wrong to Eat Your Dog?," *Journal of Personality and Social Psychology* 65, no. 4 (1993): 613–628.
2. Sara Reardon, "Yes, Your Pet Might Eat Your Corpse. That's a Problem for Investigators," *Science* 383, no. 6680 (January 19, 2024), https://www.science.org/content/article/yes-your-pet-might-eat-your-corpse-s-problem-investigators.
3. Stanley Coren, "If You Died Alone, Would Your Dog or Cat Eat You?," *Psychology Today*, February 8, 2024, https://www.psychologytoday.com/intl/blog/canine-corner/202402/if-i-die-alone-will-my-dog-or-my-cat-eat-me.
4. Reardon, "Yes, Your Pet Might Eat Your Corpse."
5. Reardon, "Yes, Your Pet Might Eat Your Corpse."

CHAPTER 22: A WEIRD RELATIONSHIP: THE ORIGIN OF DOGS, MAGNETISM, AND POO

1. B. L. Hart et al., "The Paradox of Canine Conspecific Coprophagy," *Veterinary Medicine and Science* 4, no. 2 (2018): 106–114.
2. A. Rouviere and G. Ruxton, "No Evidence for Magnetic Alignment in Domestic Dogs in Urban Parks," *Journal of Veterinary Behavior* 49 (2022): 71–74.

CHAPTER 24: DO YOU LOOK LIKE YOUR DOG?

1. Stanley Coren, "Do People Look Like Their Dogs?," *Anthrozoös* 12, no. 2 (1999): 111–114.
2. Michael M. Roy and Nicholas Christenfeld, "Do Dogs Resemble Their Owners?," *Psychological Science* 15, no. 5 (June 2004).
3. A study completed about a decade after this research identified the eyes as a crucial feature. Even when the mouths of dogs and humans were blacked out, similarities were still perceived by onlookers. Sadahiko Nakajima, "Dogs and Their Owners Resemble Each Other in the Eye Region," *Anthrozoös* 26, no. 4 (2013): 551–556.

4. Yana Bender et al., "Like Owner, Like Dog—A Systematic Review About Similarities in Dog-Human Dyads," *Personality and Individual Differences* 233 (February 2025): 112884, https://doi.org/10.1016/j.paid.2024.112884.
5. Stefan Stieger and Martin Voracek, "Not Only Dogs Resemble Their Owners, Cars Do Too," *Swiss Journal of Psychology* 73, no. 2 (2014): 111–117.

CHAPTER 25: A PSYCHIC DOG

1. Rupert Sheldrake and Pamela Smart, "A Dog That Seems to Know When His Owner Is Coming Home: Videotaped Experiments and Observations," *Journal of Scientific Exploration* 14, no. 2, (2000): 233–255.
2. Rupert Sheldrake, "Can Morphic Fields Help Explain Telepathy?," *Mindfield* 11, no. 1 (2010): 26–33.
3. Richard Wiseman, Matthew Smith, and Julie Milton, "Can Animals Detect When Their Owners Are Returning Home? An Experimental Test of the 'Psychic Pet' Phenomenon," *British Journal of Psychology* 89, no. 3 (August 1998): 453–462, https://doi.org/10.1111/j.2044-8295.1998.tb02696.x.
4. Alex Tsakiris, host, *Skeptico* (podcast), April 14, 2007, http://content.blubrry.com/skeptiko/skeptiko-2007-04-17-39683.mp3.
5. Lillian Gissen, "PET Psychic Who Charges $370 AN HOUR Reveals How She Quit Her Job as an Attorney to Become a Full-Time Animal 'Telepath' with a Waitlist of 7,600," *Daily Mail*, November 19, 2023, https://www.dailymail.co.uk/femail/article-12762185/animal-communicator-telepathy-talks-pets-mind.html.

CHAPTER 26: THE BEST-BEFORE DATE

1. "Pet Taxidermy & Pet Preservation," PreyTaxidermy.com, accessed April 14, 2025, https://www.preytaxidermy.com/pages/pet-taxidermy-services.
2. Zoom interview with Allis Markham, December 13, 2024.

CHAPTER 27: BRING 'EM BACK ALIVE

1. Per Olof Olson et al., "Insights from One Thousand Cloned Dogs," *Scientific Reports* 12, no. 11 (2022): 11209, https://doi.org/10.1038/s41598-022-15097-7.
2. Jessica Pierce, "You Love Dogs? Then Don't Clone Them," *New York Times*, March 6, 2018, https://www.nytimes.com/2018/03/06/opinion/clone-pet-streisand-dog.html.
3. Olsson et al., "Insights from One Thousand Cloned Dogs."
4. Barbra Streisand, "Barbra Streisand Explains: Why I Cloned My Dog," *New York Times*, March 2, 2018, https://www.nytimes.com/2018/03/02/style/barbra-streisand-cloned-her-dog.html.
5. Stacy Liberatore, "Would You Pay $30,000 to Freeze Your Beloved Pet . . . So You Can Be Reunited in the Future?," *Daily Mail*, May 18, 2023, https://www

.dailymail.co.uk/sciencetech/article-12091381/Company-backed-billionaire-Peter-Thiel-cryogenically-freezing-PETS-30-000.html.

CHAPTER 28: COULD THERE BE ROBOT PETS?

1. Gail F. Melson et al., "Robotic Pets in Human Lives: Implications for the Human-Animal Bond and for Human Relationships with Personified Technologies," *Journal of Social Issues* 65, no. 3 (September 2009): 545–567, https://doi.org/10.1111/j.1540-4560.2009.01613.x.
2. Peter Kahn et al., "Robotic Pets in the Lives of Preschool Children," *Interaction Studies* 7, no. 3 (January 2006): 405–436, https://doi.org/10.1075/is.7.3.13kah.
3. Emily Shoesmith et al., "Using PARO, a Robotic Seal, to Support People Living with Dementia: 'What Works' in Inpatient Dementia Care Settings?," *Human-Animal Interactions* 12, no. 1 (2024): https://doi.org/10.1079/hai.2024.0023.
4. Angela Johnston, "Robotic Seals Comfort Dementia Patients but Raise Ethical Concerns," *Crosscurrents*, KALW, August 17, 2015, https://www.kalw.org/show/crosscurrents/2015-08-17/robotic-seals-comfort-dementia-patients-but-raise-ethical-concerns.
5. Lisa Carver, "Don't Try to Replace Pets with Robots—Instead, Design Robots to Be More Like Service Animals," *The Conversation*, July 18, 2021, https://theconversation.com/dont-try-to-replace-pets-with-robots-instead-design-robots-to-be-more-like-service-animals-164522.
6. Kate Darling, *The New Breed: What Our History with Animals Reveals About Our Future with Robots* (Henry Holt, 2021).
7. Sherry Turkle, *Alone Together: Why We Expect More from Technology and Less from Each Other* (Hachette, 2017).
8. Manoush Zomorodi, Katie Monteleone, and Sanaz Meshkinpour, "If a Bot Relationship FEELS Real, Should We Care That It's Not?," *Body Electric*, NPR, July 2, 2024, https://www.npr.org/2024/07/01/1247296788/the-benefits-and-drawbacks-of-chatbot-relationships.

CHAPTER 29: TALK TO THE ANIMALS

1. Yossi Yovel and Oded Rechavi, "AI and the Doctor Doolittle Challenge," *Current Biology* 33 (2023): 783–787.

CHAPTER 30: THE WILD WORLD OF COMMUNICATION

1. "What Is the Sixth Mass Extinction and What Can We Do About It?," World Wildlife Fund, accessed April 14, 2025, https://www.worldwildlife.org/stories/what-is-the-sixth-mass-extinction-and-what-can-we-do-about-it.

2. Ryan Truscott, "Artificial Intelligence Could Soon Match Footprints to the Animals That Made Them," *Hakai Magazine*, August 12, 2024, https://hakaimagazine.com/news/artificial-intelligence-could-soon-turn-anyone-into-an-expert-tracker.
3. Aesop, "The Crow and The Pitcher," in *The Æsop for Children* (Washington, DC: Library of Congress, n.d. [1919]), https://read.gov/aesop/012.html.
4. "Clever Crow Uses Physics to Get Its Food," New Scientist, YouTube video, March 26, 2014, https://www.youtube.com/watch?v=NGaUM_OngaY (but a warning: the accompanying music is horrible).
5. "Walk on the Wild Side—Prairie Dog Alan," space-time systemz, YouTube video, August 26, 2009, https://www.youtube.com/watch?v=SNfQda8ceGs.
6. William Brennan, "The Future of Pets," *The Atlantic*, January/February 2016, https://www.theatlantic.com/magazine/archive/2016/01/the-future-of-pets/419133/.

CHAPTER 31: YOU TALKIN' TO ME?

1. A. Bastos et al., "How Do Soundboard-Trained Dogs Respond to Human Button Presses? An Investigation into Word Comprehension," *PLOS One* 19, no. 8 (2024): e0307189.
2. A. P. Rossi and C. Ades, "A Dog at the Keyboard: Using Arbitrary Signs to Communicate Requests," *Animal Cognition* 11 (2011): 329–338.
3. P. F. Cook et al., "Awake Canine fMRI Predicts Dogs' Preference for Praise vs Food," *Social Cognitive and Affective Neuroscience* 11, no. 12 (2016): 1853–1862.
4. A. Andics et al., "Neural Mechanisms for Lexical Processing in Dogs," *Science* 353, no. 6303 (2016): 1030–1032.
5. Jacob Beck, "Can We Really Know What Animals Are Thinking?" *The Conversation*, September 5, 2019, https://theconversation.com/can-we-really-know-what-animals-are-thinking-122678.
6. Temple Grandin, *Animals in Translation* (Simon & Schuster, 2006).

CHAPTER 32: A CAUTIONARY TAIL

1. E. Hobkirk and S. Twiss, "Domestication Constrains the Ability of Dogs to Convey Emotions via Facial Expressions in Comparison to Their Wolf Ancestors," *Scientific Reports* 14 (2024): 10491.
2. M. V. Kujala et al., "Empathy Enhances Decoding Accuracy of Human Neurophysiological Responses to Emotional Facial Expressions of Humans and Dogs," *Social Cognitive and Affective Neuroscience* 19, no. 1 (2024).
3. B. L. Deputte, "Heads and Tails: An Analysis of Visual Signals in Cats, *Felis catus*," *Animals* 11 (2021): 2752.

4. Tom Rottier et al., "'Tail Wags the Dog' Is Unsupported by Biomechanical Modeling of Canidae Tails Use During Terrestrial Motion," preprint, BioRxiv, December 2023, https://doi.org/10.1101/2022.12.30.522334.
5. Angelo Quaranta et al., "Asymmetric Tail-Wagging Responses by Dogs to Different Emotive Stimuli," *Current Biology* 17, no. 6 (2007): R199–201, https://doi.org/10.1016/j.cub.2007.02.008.
6. Marcello Siniscalchi et al., "Seeing Left- or Right-Asymmetric Tail Wagging Produces Different Emotional Responses in Dogs," *Current Biology* 23 (November 18, 2013): 2279–2282.
7. Wei Ren et al., "Left-Right Asymmetry and Attractor-like Dynamics of Dog's Tail Wagging During Dog-Human Interactions," *iScience* 25, no. 8 (August 2022): 104747, https://doi.org/10.1016/j.isci.2022.104747.
8. S. Leonetti et al., "Why Do Dogs Wag Their Tails?," *Biology Letters* 20 (2024): 20230407.

CHAPTER 33: FUTURE PETS

1. Ameya Paleja, "Pet Geneticists Use AI to Visualize How Dogs Will Look in 10,000 Years," *Interesting Engineering*, November 23, 2023, https://interestingengineering.com/culture/ai-visualize-dogs-in-10000-years.
2. Emily Mullin, "Your Next Pet Could Be a Glowing Rabbit," *Wired*, February 19, 2025, https://www.wired.com/story/your-next-pet-could-be-a-glowing-rabbit-los-angeles-project-gene-editing-crispr/.
3. Ashley Kilroy, "Adoption Statistics 2024," *Forbes*, January 2, 2025, https://www.forbes.com/advisor/pet-insurance/pet-adoption-statistics/.
4. C. Kraus et al., "The Size–Life Span Trade-Off Decomposed: Why Large Dogs Die Young," *American Naturalist* 181, no. 4 (2013): 492–505.
5. Cat Healthy Aging Project, accessed April 14, 2025, https://cathealthyagingproject.org/f-a-q/.
6. Jonathan Losos, "House Cats Will Rule the World," *Slate*, October 23, 2023, https://slate.com/technology/2023/10/house-cats-evolution-anthropocene-extinction.html.
7. Jessica Pierce and Marc Bekoff, *A Dog's World: Imagining the Lives of Dogs in a World Without Humans* (Princeton University Press, 2021).
8. Mark Bekoff, "As Dogs Go Wild in a World Without Us, How Might They Cope?," *Psychology Today*, September 2, 2018, https://www.psychologytoday.com/ca/blog/animal-emotions/201809/dogs-go-wild-in-world-without-us-how-might-they-cope.
9. Jonathan Chadwick, "Meet Goldilocks, the Breedless Dog of the Future: All Breeds Would Merge into ONE Within Just Five Years Without Humans, Expert Claims—Here's What It Would Look Like," *Daily Mail*, April 6, 2024, https://www.dailymail.co.uk/sciencetech/article-13268285/breedless-dog-merge-humans.html.

10. G. J. Spatola et al., "The Dogs of Chernobyl: Demographic Insights into Populations Inhabiting the Nuclear Exclusion Zone," *Science Advances* 9 (2023).
11. Patrick Pester, "Could Dogs Survive Without Humans?," *LiveScience*, February 25, 2023, https://www.livescience.com/could-dogs-survive-without-humans.

CONCLUSION

1. Yi-Fu Tuan, *Dominance and Affection: The Making of Pets* (Yale University Press, 1984).

About the Author

Jay Ingram has hosted two national science programs in Canada, *Quirks & Quarks* on CBC Radio and *Daily Planet* on Discovery Channel Canada. He is the author of twenty books, which have been translated into fifteen languages, including the bestselling five-volume *Science of Why* series. In 2015, he won the Walter C. Alvarez Award from the American Medical Writers Association for excellence in communicating health care developments and concepts to the public, and from 2005 to 2015 he chaired the Science Communications Program at the Banff Centre. Ingram has seven honorary degrees, was awarded the Queen Elizabeth II Diamond Jubilee Medal, and is a Member of the Order of Canada. He is a cofounder of the Beakerhead arts and engineering street festival in Calgary and lives in Victoria, British Columbia.